# Frames and Fictions
# on Television

## The politics of identity within drama

Edited by
## Bruce Carson and
## Margaret Llewellyn-Jones

D1354462

**intellect** TM
EXETER, ENGLAND
PORTLAND, OR, USA

First Published in Paperback in 2000 by
**Intellect Books**, FAE, Earl Richards Road North, Exeter EX2 6AS, UK

First Published in USA in 2000 by
**Intellect Books**, ISBS, 5804 N.E. Hassalo St, Portland, Oregon 97213-3644, USA

| | |
|---|---|
| Publisher: | Robin Beecroft |
| Cover Design: | Bettina Newman |
| Copy Editor: | Lucy Kind |

A catalogue record for this book is available from the British Library

ISBN 1-84150-009-7 (cloth)
ISBN 1-84150-050-X (paper)

Printed and bound in Great Britain by Cromwell Press, Wiltshire

# Contents

# Framing and Re-Framing the 'Other'

# Introduction: Issues of Cultural Identity

## Bruce Carson and Margaret Llewellyn-Jones

*Frames and Fictions on Television: The Politics of Identity within Drama* is a collection of essays whose contributors use a variety of approaches to explore how issues of cultural identity have been mediated through contemporary TV Drama. At a general level issues of identity and difference have become shorthand ways of describing current debates within Cultural Studies about the nature and impact of recent economic, political and social changes within capitalism. A variety of complex and contradictory sources have been seen as contributing to this, for example the long-term influence of social movements such as feminism, the re-emergence or re-negotiation of ethnic and national identities in a post-Cold War era, changing definitions of personal and sexual identities and most significantly the 1980s political resurgence of the free-market philosophy of the New Right that became the ideological justification for the accelerating pace of capitalist modernisation or what is termed in media discourses as 'globalisation'.

For some theorists[1] these economic and cultural developments have created a range of concerns that can be variously traced across the contributions to this volume. Paramount among these is the feeling that in the 1990s identity is becoming more mobile and subject to change and innovation. This emphasis on the individual, rather than overt ideological and communal concerns, can be linked back to the zeitgeist mood of the 1980s represented by Mrs Thatcher's notorious statement, 'There is no such thing as society'. It is crucially in the areas of culture that these transitions are most evident. The media industry being an important mediating force in the way that individuals make sense of their own lives and identities. In the forefront of these changes is television, a major industry in the development of a transnational media culture. The expansion of the global media market and the rise of new information technologies over the last 30 years has strong implications, both now and in the future, for all national and regional cultural identities, TV audiences, their reading practices and programme formats. This is echoed in the increased volume of TV material to be read, as well as the consumer's ability to control strategies and timing of their viewing that have evolved into fragmenting the mass audience into a plethora of different and multiple reading positions. It is thus difficult to ascertain whether the bardic function claimed for television by Fiske and Hartley[2] is still tenable. The audience which in general terms is more 'knowing' about the processes of TV and the way it has been constructed, is composed of individuals whose identity is both more fluid and dislocated.

## Fears of Mass Culture

A key fear about the globalisation of culture[3] is that along with the threat to national broadcasting cultures, there is a growing convergence and homogeneity in cultural representations under the influence of transnational corporations driven by profit and the new digitised technologies to reach the largest number of consumers in the global market place. Global corporations like Disney, Time Warner, Sony, Viacom, and News Corporation are not only expanding their control over programming, distribution and transmission systems, but are seen as aiming to move beyond an era of mass media into one that is geared to personalised media and individual choice. There have long been fears and resentment in Britain and Europe about cultural decline and loss of national identity brought about by the threat from American mass culture. In the 1950s the threat of Americanisation was best expressed by moral conservatives' fears about the bad influence of milk bars and American rock n'roll on British youth (Hebdige, 1988: 56). In the 1990s this resentment of American media dominance has been recently expressed on a European-wide basis through GATT trade negotiations in a variety of markets, including the cinema with its fears of the 'Hollywoodisation' of Europe. However, there is another side to US corporate dominance of national media markets in that its commercial processes are not always just one-way, they can also be shaped by cultural difference, viz. the social structures and ideologies of existing national cultures. For example, in the mythical 'swinging sixties' a burgeoning British pop music and youth culture was successfully re-exported back to America and other overseas markets.

In media terms these fears are an extension of a twentieth century mass society tradition that sees the mass of 'the people' in modern society as alienated, ill-educated and rootless, cut off from a sense of organic community and open to media manipulation. Such fears were again expressed in 1983 by the BBC's Director-General, Alisdair Milne, with his famous phrase 'wall-to-wall Dallas'. His commentary on this challenge to our national culture showed his concerns about American dominance of the international television industry and the impact American programmes would have on British audiences; the popular soap opera *Dallas* being the most visible evidence of such a cultural invasion in the early 1980s. Underlying these cultural prejudices about programming quality lay deep concerns in the BBC about the post-1979 Tory government's plans to change the identity of British television by opening it up to greater competition. These debates about programming quality have always masked underlying anxieties about threats to national and regional culture and related ideas about moral standards. They have a long history which needs to be explored before we can discuss the current identity of British television and the status of its dramatic output.

## A National Broadcasting Culture

Television itself has not been immune to changes over the last 30 years. Since the mid-1980s, the UK television industry has steadily moved from a public service model to one increasingly based upon market forces. In the process, our national broadcasting culture has become increasingly fragmented. In the 1960s its identity was more clear-cut, based as it was around a State-influenced BBC/ITV duopoly that, even after the advent of commercial TV in 1955, still operated with a Reithian public service ethos of social purpose and moral responsibility. Despite the creation of regional ITV companies this structure was still heavily focused on London and continued to act as a powerful national unifying force operating through a broadcasting culture that offered a diversity of programming. Television was seen as having a duty to educate and inform its mass audience. Indeed, within the hierarchy of television itself, drama was a key area for inculcating artistic and literary values within a family medium that was perceived by many critics as having little cultural value.

From the late 1950s onwards there was a flowering of plays written for television that coincided with the cultural renaissance in literature and the theatre. This televisual 'New Wave' was undoubtedly helped along by the changes in the organisation and planning brought about through the impact of commercial television on the BBC monopoly in the mid-1950s. Institutional spaces opened up within ITV, and later the BBC, for experimentation which were also encouraged by the 1962 Pilkington Committee's report 'that it is television's moral responsibility to shun triviality and risk challenge and controversy' (Caughie, 1996: 218). Indeed, ITV's *Armchair Theatre* (created by ABC Television in 1956) had already been modernised through the recruitment in 1958 of a Canadian TV executive, Sydney Newman. The subsequent popularity of this live drama series, scheduled as it was after *Sunday Night at the London Palladium*, resulted in Newman being head-hunted by the BBC in 1963 to reorganise its own Drama department. This led to the launch in 1964 of what soon became the quintessential drama slot of 1960s television, *The Wednesday Play* (renamed *Play for Today* in 1968). These programme areas expanded opportunities for new writers to deal with contemporary issues and TV drama often became the focus of intense political debate. There developed a strong sense amongst some critics and practitioners that the 'authored' single play offered the best area in television for giving space to alternative or oppositional voices and identities. It also offered a space for new writers to experiment. This was to be a 'progressive' TV drama that at its best challenged the dominant artistic and political conventions of the day. The desire to experiment was also aided by new lightweight 16mm film technology that helped dramatists and film-makers to escape the confines of the TV studio. Such developments gave rise initially to the greater influence of documentary forms, as exemplified by the controversial 1966 drama-documentaries *Cathy Come Home* and *The War Game*. What has happened since this era has been a gradual closing down of this institutional space for radical drama as a result of changes to the political economy of television in the last 30 years.

## An International Market Place

Left-wing/liberal commentators often forgot that there was also a great deal of popular mainstream naturalistic drama. The success in the 1960s of long-running drama serials like *Coronation Street* and *Emergency Ward Ten*, drama series like *Dr Finlay's Casebook*, *The Power Game*, *The Troubleshooters*, *Z Cars*, and situation comedies like *The Likely Lads*, *Steptoe and Son*, and *'Till Death Us Do Part* showed that the home industry could produce quality entertainment for large home audiences. However there was already overseas competition in the shape of US television companies that had developed as successful offshoots of the Hollywood studios or 'majors' in the early 1950s. Despite restrictions on the amount of foreign imports entering the UK, the success with home audiences in the 1950s/1960s of programmes like *Bonanza*, *The*

*Fugitive, Dr Kildare, I Love Lucy, The Phil Silvers Show, The Untouchables, The Man From Uncle* and *Peyton Place*, was an early glimpse of a transnational television culture. This trend has continued up to the present day with the international success of American prime-time soaps like *Dallas* and *Dynasty*, through *Miami Vice* to more recent dramas like the much lampooned but heavily 'watched' *Baywatch*. This is not unsurprising given that Hollywood had been producing genre films for years that had crossed national boundaries and cultural identities in their desire to win large international audiences and dominate national cinemas. However, the arrival of a more business orientated ITV in 1950s Britain saw the gradual use of film recording and the new video technology by the early 1960s. For example, the new ITV not only purchased American programmes but started to sell drama to the lucrative US market. The first example, being the filmed children's drama series *The Adventures of Robin Hood* (produced by Sapphire Films) which was initially sold to the US in 1958 before its transmission in the UK. Its early success in the US was later emulated in the 1960s by spy-influenced 'pop' dramas for ITV like *The Avengers* (ABC Television, 1961-69) and *The Prisoner* (ITC/Everyman Films, 1967-68)

Thus, well before the seismic political shifts to the right in the 1980s the commodification of television was proceeding apace under the impact of competition on its long-established public service culture. Such changes also made it more difficult by the late 1970s to justify prestigious single dramas. The British television industry was coming under the sway of an economic logic that preferred serial or series formats whose initial high costs were offset by the advantage of repeatability. Moreover, it was increasingly looking beyond its small national audience to the prestige and profits of the international market place. For example, Thames TV, through its subsidiary Thames Television International (TTI) became the most successful ITV company in foreign markets in the 1970s on the basis of its sitcoms, such as *George and Mildred, Man About the House* and *Robin's Nest*, light entertainment shows like *Benny Hill* and classic serial dramas such as *Edward and Mrs Simpson*. Thames TV had also set up its own film company, Euston Films Ltd in 1971 (Alvarado and Stewart, 1985: 12) which had contributed to TTI's export success by producing several successful drama series in the 1970s/1980s that had a very 'British' flavour, for example *The Flame Trees of Thika, Minder, Out, Reilly – Ace of Spies, Special Branch, The Sweeney* and *Widows*. Their strong 'British' identity helped many of these series to do well in terms of foreign sales (Alvarado and Stewart, 1985: 12). Another way to achieve international success was to finance the higher costs of culturally prestigious 'classic serials' through co-production deals (Kerr, 1982: 18) that cut across national boundaries and appealed to international audiences. These International style dramas had a politically safer subject matter that was mainly based either on figures or families of historical importance or were literary adaptations of work by 'classic' authors. The huge world-wide success of the BBC's first major 'heritage-export' drama *The Forsyte Saga* in 1970 accelerated a trend that has continued to see British film and TV companies use nostalgic evocations of national culture and identity as a way of competing 'head-on' at home and abroad with US films and 'made for TV' film imports.

## TV and TV Drama in the 1980s: Competition and De-regulation

These market-led processes were accelerated from the mid-1980s onwards by the Thatcherite zeal of a reforming Tory government which unleashed a critical debate over the future shape of British television. It was a debate that ultimately led to the Broadcasting Act of 1990 and a further shift away from the public to the private sphere. The subsequent moves towards industry deregulation have seen a further fragmentation of the national broadcasting culture in the face of a global television market place dominated by US production. This rearticulation of the relationship between the global, the national and the local has seen cable and satellite multi-channel companies like BSkyB push aggressively for new 'global' markets from the late 1980s onwards. These changes have intensified in the 1990s, and look set to develop further through the growing technological impact of digital and internet technology on television in the twenty-first century. At the same time there has been a restructuring of the industry with a greater concentration of ownership both nationally and internationally, higher levels of independent production and more imported TV programmes.

## Quality TV Debates

A central issue in the debate about the proposed changes in British television was programming quality. In the late 1980s, the term 'quality TV' became much talked about in that it was used both by those pressure groups who supported the opening up of a duopolistic industry to greater competition, as well as by those who were anxious to protect programming diversity and standards by retaining the principles of public service broadcasting. At the centre of this debate was the identity of British TV with its world-wide reputation for high quality programmes, an identity that according to Charlotte Brunsdon was exemplified in the 1980s by classic serials like Granada TV's *Brideshead Revisited (1981)* and *The Jewel in the Crown (1984)*. Such historical dramas are rooted in British literary and theatrical traditions and, as Brunsdon argues (Brunsdon, 1990: 84), regularly bring plaudits to home television. This commercial strategy has continued to be important to the industry in the 1990s, as can be seen with the recent successful adaptations of George Eliot's *Middlemarch* (BBC2, 1994), and Charles Dicken's *Our Mutual Friend* (BBC2, 1998), as well as perennial favourites like Jane Austen's *Pride and Prejudice* (BBC1, 1995). However, these were not the only examples of how the identity of TV fiction was changing from an earlier era.

## Channel 4 and Drama

The 1980s saw the tradition of the 'authored' single play not disappear entirely but instead alter direction. There were still 'serious' and successful dramas being made that were critical of the status quo. Whilst fewer in number, these dramas were increasingly likely to use a serial format, for example *Boys from the Blackstuff, Edge of Darkness, The Monocled Mutineer, The Singing Detective* and *A Very British Coup*. However, the advent of a new television channel (Channel 4) in November 1982, saw another important shift in the output of TV fiction. As John Hill has pointed out, Channel 4 introduced a more 'flexible, post-Fordist mode of production' (Hill, 1996: 156) into British television. The new channel operated on a publishing house model and commissioned programmes from a variety of production companies. Their programming policy had a public service remit in that the channel had to appeal to previously ignored tastes and audiences. In practice this saw the channel's schedulers trying to increasingly attract a younger, more affluent and media-literate audience for advertisers. The channel's influence was particularly important in two areas of drama:

## a) Film on Four

The first of these changes was the introduction of low-budget TV funded films under the *Film on Four* banner. It had a hybrid approach that was already operated in other countries, such as the regionally based German TV companies like *ZDF* and *WDR*. *Film on Four* created an institutional space for both genre film-making as well as films that combined elements of art-house cinema with the earlier socio-political concerns of naturalist/realist TV drama. Eventually the new channel developed a successful strategy of cinematic release followed by a later television launch. Such films as *The Draughtsman's Contract, Distant Voices, Still Lives, The Ploughman's Lunch, My Beautiful Laundrette, Mona Lisa, Gregory's Girl, Letter to Brezhnev, The Crying Game, Riff-Raff* and *Naked* explored a wide range of issues and cultural identities in a period of political conservatism. Although films like these rarely break into even the top seventy list of mainstream TV shows their attributes are important in that they are culturally and politically seen as high quality drama. They have gained both success and prestige for the new channel whilst giving a boost to the home film industry in the 1980s and 1990s thus intensifying the aesthetic and economic relationship between British film companies and television. In fact *Film on Four* represented such a successful approach in terms of viewing figures and cultural prestige that it encouraged other ITV companies like Central and Granada Television to set up their own film-making subsidiaries and the BBC with its *Screen 2* format to follow a similar commercial strategy of making films intended for cinema release (Hill, 1996: 162-3).

## b) Quality American Drama

Another trend has been the scheduling of 'quality' programming from America. This approach was pioneered in the 1980s by MTM Enterprises Inc., an independent production company, with such politically liberal and popular drama series such as, *St Elsewhere, Lou Grant* and the innovative *Hill Street Blues*. Its flexi-narrative approach of 'multi-narrative, multi-character television drama was common in soaps by the 1970s' (Nelson, 1997: 30), but not in drama series. It has been so much copied that the flexi-narrative has become an industry norm. The 1980s also saw the import of other quality US comedies and drama series such as *Cagney and Lacey, Cheers, LA Law, Thirtysomething* and the formally innovative *Moonlighting*. This is a trend that has been continued into the 1990s with Channel 4 and the BBC buying American programmes like *ER, Frasier, Friends, Homicide, Murder One, NYPD Blue, Seinfeld, The Larry Sanders Show* and *The X-Files*. In addition, Channel 4 helped to pioneer a trend in repeating 'classic' British and American comedies and dramas from earlier periods that also appealed to younger and more media-literate viewers; another Channel 4 policy that has also been subsequently copied by other channels.

## TV Drama and Soap Opera

Another way for TV companies to spread the high programme costs of drama has been through the development of soap operas. These are almost as old as television itself, but in Britain really came into their own with television's first soap opera *The Grove Family* (1954–57), a medical soap *Emergency Ward Ten* (1957–67), *Coronation Street* (1960 to present), *Crossroads* (1964–88), and *Emmerdale Farm* (now *Emmerdale*) from 1972 onwards. The later international success of the more melodramatic prime-time American soap operas like *Dallas* and *Dynasty* in the late 1970s was not surprising given the American industry's experience at creating films and TV programmes with a transnational cultural appeal. However, British soaps have remained more parochial with their roots in a realist aesthetic with its emphasis on social issues, authentic locations and regional cultures. The success of Channel 4's *Brookside* was based on changes to the home product that placed it in a 1980s style suburban location. It was miles away from the cosy nostalgia of *Coronation Street's* northern, working class community and appealed to a new and younger generation of viewers.

The BBC had always wanted a rival soap opera to compete with the long-running success of ITVs *Coronation Street.* In the early 1960s it tried and failed with its own soap, *Compact* (1962–65). The Corporation has always had contradictory requirements in that it must win a large enough audience share to justify its licence fee to its political masters, whilst at the same time invest in less popular programmes that will support its public service ethos. With the pegging of the licence fee by a politically unsympathetic Conservative government in the 1980s it needed to find audiences in order to justify its existence. The success of *EastEnders* (1985 to present) was part of its strategy to continue competing in the mainstream of television entertainment.

As a dramatic form, the cliffhanger endings, multiple characters and storylines have influenced drama series across a range of genres, from police dramas like *Hill Street Blues* and *The Bill* to hospital dramas like *Casualty* and *ER*. If it can be argued that the drama-documentary *Cathy Come Home* was a national event in 1966, then the power of soap opera was seen in the 1993 story about *Brookside's* Jordache family. This complex narrative of rape and domestic violence, experienced respectively by Beth Jordache and her mother Mandy, continued the social issue tradition but also helped to restore the programme's flagging audience figures. The subsequent murder and accidental discover of Trevor Jordache's body by Sinbad Sweeney under the next door neighbour's patio was a discovery that was kept secret for eighteen months until the mother and daughter's murder trial in 1995. Such has been the success in winning audiences that soap opera has become the mainstream drama at both a national and international level.

## Television Drama in the 1990s

The current realities that help to shape television scheduling in the 1990s are an increasing level of competition between terrestrial, cable and satellite companies for the TV viewer. With no sign that the average adult weekly viewing figure of around 27 hours per person is likely to increase, programme executives are faced with an audience that is fragmenting under the new competitive pressures. The majority of home-made television drama programmes are made as series or serials, with the top twenty programmes consistently filled by soap operas, situation comedies and a new phenomenon, the prime-time drama. British TV drama productions in the 1990s have been dominated by dramas like *Cracker, Inspector Morse, Peak Practice, Prime Suspect, Taggart, Heartbeat, The Darling Buds of May* and *London's Burning*. An area of drama in which the BBC has consistently lagged behind the ITV companies, with *Casualty* being one of its few successes.

With single plays having virtually disappeared this means that, with a few exceptions, very little drama is made that is formally or politically challenging to the status quo or deals with minority or under-represented groups in the 1990s. The fact that soap operas like *Brookside* and *EastEnders* are part of a realist tradition that explores contemporary social issues also has to be put within the context of their focus on the personal and the emotional (Geraghty, 1991: 41) rather than the wider social and historical factors that shape relationships between individuals. Undoubtedly the culture and politics of the 1980s and 1990s has seen a decline in drama terms from a politics aimed at increasing audience knowledge to one that is a politics of pleasurable consumption. This is shown by the fact that there is a huge convergence of top-rated drama towards the values of entertainment. The consequent debate that has ocurred on 'quality' mirrors many of the fears and anxieties expressed in the run up to the 1990 Broadcasting Act. These include charges in the broadsheet and tabloid newspapers that the BBC is 'dumbing down' and moving away from its Reithian policies of the past and that Channel 4 is abandoning its mission. For example, Alison Pearson, a journalist who regularly appeared on the BBC's *Late Review*, nevertheless wrote in her London *Evening Standard* column (July 25 1999) that the BBC were simply chasing audience figures at the expense of standards. Under criticism from the BBC Board of Governors and with Greg Dyke newly appointed as Governor General, Peter Salmon defended himself from the criticism of aping ITV's formats with the claim that they were scheduling three or four returning drama series and writer-led dramas (*The Guardian*, July 5 1999). Interestingly, three of these writers were successful women – Lucy Gannon, Debbie Horsfield and Paula Milne – whilst the other, Andrew Davies has been responsible for several 'quality' adaptations. Brian Park (*The Guardian*, 3 August 1999) gives the cost of soap episodes at between £45-90,000, prime-time drama such as *Inspector Morse* as £600-700,000, and period drama as £1 million and over. Since increasingly drama tends to be made less in-house and more 'bought-in' by independent producers, economies of scale are important even in so-called quality drama. Gub Neal, who as Head of Drama at *Channel 4* has worked on innovatory programmes such as *Queer as Folk* is reported as conceding that any new drama:

> ...must conform to previously successful templates to stand much chance of being commissioned...but the trick lies in finding ways to play the system at its own game... There could be more original ideas driving the drama that's on, but ITV is a commercial machine and the BBC is driven by its remit to cater for volume rather than diversity, which means it has to juggle with some very complex imperatives...In the end, the whole thing is basically a lottery. (*The Guardian*, May 4 1999)

However, these arguments can only be properly understood in the context of increasing industry pressures to compete more effectively on an international scale. The current tensions between British television's highly regulated home market, developed largely under a public service broadcasting culture and global competitive pressures are no more evident than in the recent Government and Industry commissioned report on British television in overseas markets. The report is based on international industry statistics and selected interviews (number and type unspecified) with overseas programming executives in 12 countries. It puts forward a range of recommendations for encouraging more British television exports but also recognises that ultimate success will depend on changes in the style and content of British programming so as to more adequately meet the needs of overseas markets. This is the language of marketing and market-driven programming where focus groups and rebranding are seen as the key to future commercial success in a global market. For example, in its section on programme quality, the report concludes that 'the image of 'Britain' portrayed overseas by our television drama is not an attractive one' (David Graham and Associates, 1999: 24). It would appear that despite a strong reputation for high production values in areas like costume drama, much of our drama is still perceived by overseas TV executives as too dark and realistic for their mainstream audiences. Dramas like *Band of Gold* and *Prime Suspect* are seen as being filled with distasteful characters and storylines that are too parochially British and show a country that is a 'relatively poor, down-at-heel place which does not inspire interest' (ibid: 25). The report goes on to comment that this image does not fit one of Britain as a modern creative country. Despite some UK successes like *Cracker, Bugs, The Sculptress* and *Mr Bean*, it concludes that British TV executives 'have not yet capitalised on the rejuvenation of the British 'brand' to produce a kind of positive, glossy, mainstream drama series that would command interest overseas. The series that made Britain swing in the sixties – *The Avengers, The Prisoner, The Saint, The Champions* – have not been replaced in the nineties' (ibid:26). A similar quality problem is also felt to operate in comedy where the parochial nature and short runs of many of our situation comedies are not seem as matching the successes of the best longer run American team-written comedies. One of the report's main recommendations is a move to a more US-style market-led model of television programme making that involves making programme pilots to test out market reactions before committing the makers to long running comedy and drama series rather than the short-run tendency that is seen as current amongst most British producers.

## TV Audiences, Genres and Cultural Identity

These current struggles over the identity and structure of our television industry and the export potential of its programmes are linked to wider political currents within New Labour to rebrand Britain internationally as a modern and innovative country for the 21st century (Leonard, 1997: 8). The global market forces, institutional changes and new technology indicated above are also having an impact not just on national broadcasting structures and issues of national identity but on audiences and their programme formats. Whilst the changing focus and forms of television drama tend to privilege the private and the personal over the public and the political, an increasing mixture of genres draws upon the notion of a more active and responsive reader. A reader who enjoys both the predictive pleasures of some genres, and intertextual echoes in programmes which play with tensions between different generic conventions. Such changes also impact on the ways in which both the blurring of genres, and institutional practices such as the re-framing and targeted marketing of programmes can affect the identities of both image and spectator. In this sense, it is possible to infer a symbiotic relationship between the impact of global market forces not just on national broadcasting structures, but on the hybridisation of both genres and cultural identities, a case implied in several chapters.

Underlying attitudes to the increasing fluidity of genre is the ongoing debate about the nature of television realism as explored by Fiske and Hartley but later refined by Tulloch[4] and Nelson[5]. Several contributors allude to the potential ideological conservatism of realism, and ponder whether the potentially radical element within the social realism of the so-called 'Golden Age of TV Drama' in the 1960s can in any way linger within newer variations such as 'progressive realism' or 'critical realism', whether popular series inevitably fall into 'formulaic realism', or if there are subversive possibilities in the development of excess and carnivalesque qualities. The slippage between the 'fact' of the everyday and fiction in the increasingly popular Docudrama is particularly pertinent to the representation of identity, since such shows revel in the construction of the self as performer within typical social and work contexts. TV performance as a confirmation of existence, rather similarly underpins the proliferation of professional 'make-over' programmes where the identity of an everyday home or garden undergoes a fantasy of transformation which is structured like a drama. Gail Coles' chapter on Docusoap and Jonathan Bignell's chapter which contrasts the televisual representations of Mrs Thatcher and Princess Diana differently approach the question of the potential gendering of documentary and drama, or melodrama, as masculine and feminine respectively. Coles' account emphasises the centrality of the subject, the narrativisation of the material towards resolution and the privileging of the personal over political issues. Although the popularity of the 'ordinary people', as 'star' participants in *Driving School* and *Airport*, does not necessarily support more detached readings; as Coles indicates, the visibility of the camera while perhaps working against audience voyeurism does contribute to an excessive performance style and metatheatrical qualities which perhaps provoke potentially ironic readings for a media literate audience. Nevertheless, the centrality of emotional response for the majority, combined with the stereotypical range of 'cast' has provoked charges of 'dumbing down' from critics. Bignell's analysis of the hybrid blending of documentary drama and melodramatic forms in so-called 'factual' programmes about two iconic women also highlights the processes of narrativisation as well as the way that a focus on the emotional is in tension with the discourse of current affairs, such that the contradictions inherent in contemporary representations of the female, especially those in the public eye, are exposed. Placing these dramas in the context of other programmes and current affairs, the chapter explores their depiction of women failing to retain their power within masculine elites, and the ambiguous nature of the media's role in relation to these events and the construction of the public identity of both Diana and Mrs Thatcher.

# Introduction: Issues of Cultural Identity

Madeleine MacMurraugh-Kavanagh's chapter on *The Cops*, for which Tony Garnett was Executive Producer, shows how the ideology of identity is crucial to a radical re-writing of the police drama, such that the unsettled audience is disorientated and prompted to question the nature of police institutions. MacMurraugh-Kavanagh, through detailed analysis of televisual strategies, and with reference to an interview with Garnett, shows how this extension of the potential of the social realist drama aimed to create a dramatic representation of the invisible landscapes of deprivation beyond the image of Britain under Blair's New Labour. It also developed the approach of drama-documentary and even docusoap in bridging the gap between fact and fiction. Her discussion shows how although the conventions of narrative are disrupted, the use of handheld camera and other documentary techniques contribute to a sense of unmediat-ed reality. The subversion of the traditional notion that the criminal groups of society are 'Other' is challenged by the gretaer amount of serious crime committed by the police, and the blurring of the distinction between 'us' and 'them'. The representations of the cops themselves are ambiguous, the viewer is forced, according to MacMurraugh-Kavanagh by a kind of quasi-Brechtian alienation effect, to confront the issue that the television characters are ideologically constructed, and so is he/she. This collapse of certainty can in many ways be seen as a postmodern feature, and the slipperines of moral identity in *The Cops* is echoed in the increasing instability in the representation of gender identity which is of major interest in this book.

The centrality of gender identity to television drama, both as content and as poten-tially linked to aspects of form, is evident in several chapters, particularly those written by Deborah Philips, Robin Nelson, Jeremy Ridgman and Stephen Farrier. Both Philips and Nelson consider whether the visibility of women in TV drama may have increased, and if so, whether this is a progressive shift. Whilst their articles both explore the extent to which, despite the predominance of masculine narratives, the professional woman as doctor, pathologist or combined medical expert/investigative government agent has emerged as central protagonist, their approach and conclusions differ. Philips traces in detail the history of popular medical dramas in both England and America from the 1950s and 1960s, comparing the idealised representation of male doctors and female nurses as models for contemporary gender roles and power relations, where the hospi-tal setting was, in Britain, also originally a symbol of the health of the Welfare State. She discusses the importance of the generic mixture of soap and sometimes situation comedy elements which flavour the different series to particular effect. Citing the emergence of the relatively few 'Women Doctors' as key figures in recent American and British medical dramas such as *Casualty* and *ER*, both with ensemble cast, as typical of relevance dramas, Philips suggests that their identity is represented as flawed and ambivalent. There is often a tension between their private and professional roles, with difficulties symboli-cally displaced through literal or metaphorical kinds of incapacity. Having indicated the significance of the debate about strategies through which representations of women may both subvert the male gaze and disrupt classic narrative form, Nelson is concerned with a possible shift in the representation of professional women, which must challenge tradi-tional notions of male and female gender identity through new tactics. His three chosen roles, Dr. Sam Ryan from *Silent Witness*, Dana Scully from *The X-Files* and the doctor

11

from the Volvo advertisement, are all independent women, professionals who are in command of the rational scientific discourses usually ascribed to men. The work of pathologist as performed by Ryan even allows her the opportunity to gaze on and have ultimate power over the dead male body. After reference to earlier programmes such as the American *Cagney and Lacey* which provoked controversy in showing tough successful women in a man's world, Nelson points out that the participation of the three selected characters in the hybrid detective/medical genre can be associated with a blurring of binary oppositions of maleness and femaleness. Detailed examination of the programmes suggests to Nelson that these three protagonists are 'performing professional maleness', which may seem in contradiction with the way their appearance is encoded as 'feminine'. Like Philips, Nelson has some reservations about aspects of masculine culture which still underly the representation of professional women on television, but he suggests that the 'drip feed' impact may have a positive psychological influence in terms of empowerment upon the audience, despite the reluctance of the largely male-dominated TV institutions to produce more adventurous images of social change.

Changes in attitudes to gender are central to the chapters of Ridgman and Farrier, which analyse respectively *Our Friends in the North* and soap opera both of which could be considered to be largely realist rather than radical in form. Whether or not the emergence of feminism has provoked the crisis in masculine identity which is a central element in Ridgman's analysis, it is certain that the struggle to produce Peter Flannery's *Our Friends in the North* which seems to have much in common with the social realism of the 1970s is indicative of institutional caution, whereas soap is a television staple. Unlike soap, the high cost production values of this serial is more typical of the costume dramas often used as an indicator of quality, as Ridgman indicates. However, he sees *Our Friends* as an example of the ways in which social realism has become more subject to the interplay of intertextual discourses and popular narrative pleasures. It has a nostalgic edge which he claims binds the spectator to the events through various strategies, so that it frames political history as a drama of personal and emotional relations. It is thus indicative of the shift from the public to the personal indicated above. Having indicated both production history and the changes made from the original stage version, Ridgman concentrates on the increasingly privatised scenario that he claims links sexual and political identities in a drama which enacts the crisis of masculinity. This is done through the representation of the three main male friends Nicky, Geordie and Tosker, whom he considers that the narrative favours over the female, Mary. Nick's progress from political activist to successful photographer of revolutionary images is analysed in detail in the context of a Lacanian approach to subjectivity which also interprets domestic and personal struggle through the Law of the Father, whilst brief reference is made to other political dramas.

Whereas it might be said that the success of *Our Friends* depends on the notion of bardic function in that its nostalgic pleasures depend upon the audience seeing their recent past reflected back, Farrier's chapter tackles the extent to which soap can claim to have bardic function in the face of marginalised, exnominated readers, whose identities are not sited a position of socio-centrality. In a complex discussion which compares the characteristics of both queer and soap, and with reference to *EastEnders* and other soaps, he examines the ways in which the representation of gay characters in soaps is subject to a clawback to centrality. Presumably, in other words, Harold/Hayley, the trans-sexual in *Coronation Street* gradually becomes 'one of the girls'. In Farrier's terms, such clawback works against the effective representation of queers in Soap. After posing questions about the possibility of queer reading practice, with reference to the kind of looking required by Holbein's picture *The Ambassadors* and to Gerbner's model of communication he proposes a new model for queer reading which allows for the ludic qualities of both soap and queer.

The notion of making the invisible visible which is associated with the idea of queer reading practices, is also an issue with the chapters written by Claire Tylee, Bruce Carson and Margaret Llewellyn-Jones which are concerned with ethnic and national identity. The representation of 'Otherness' is ideologically complex in a multi-racial society with a post-colonial heritage, where a variety of reading positions within the broad audience may either be exnominated, or may be confronted with offensive stereotyping. Janet MacCabe on the other hand is concerned with the way that British television institutions re-frame quality programmes made in America, thus producing new identities from the 'Other' culture. These chapters either share overtly or implicitly notions about the potential of hybrid identities.

Tylee's chapter centres upon three works by black women dramatists, all transmitted in 1992, which subvert dominant generic conventions and perpetuate the silence, invisibility or stereotyping of Black women. Meera Syal, Jackie Kay and Winsome Pinnock all graduated from University in 1982/1983, have worked successfully in theatre and have moved from fringe to mainstream. Syal was born near Wolverhampton of South Asian parents, while Kay was born in Edinburgh of a Scottish mother and African father, but adopted and raised in Glasgow. Pinnock was born in Islington to Jamaican parents. As a white academic, Tylee is appropriately tentative, drawing upon self-definitions offered by Black British feminists, whilst questioning the scarce opportunities for such writers in television. Indicating the intimate knowledge of plural cultures experienced by these similar yet different writers, Tylee stresses the creative possibilities of such hybrid identities, reflected in the ways that the three works cited all feature Black, liminal protagonists whose exploration of an unknown and hidden culture facilitates questioning of the dominant culture from which they embark in their quest. The audience is seduced to accompany these characters on a border-crossing journey, which, according to Tylee, causes them not to look not only at this 'Other', but also in a new way at themselves. Tylee compares the three works in some detail, suggesting that the dramatic strategies used are successful to different degrees, and vary in the extent to which they suggest that it is possible to embrace an 'Other' across differences in culture and race.

Carson's chapter, like that of Llewellyn-Jones, places the 'Other' more overtly within the post-colonial context, linking notions of cultural and generic hybridity, and acknowledging the effect of the Asian and Irish diasporic television audience in Britain. The 1993 adaptation of Hanif Kureishi's novel for television draws upon a nostalgia for the recent historic past that is linked by Carson to the institutional pressures which earlier encouraged Channel 4 to commission work from independent film-makers from racial and ethnic minority backgrounds. Nevertheless, Kureishi's earlier films, most notably *My Beautiful Laundrette*, directed by Stephen Frears, provoked strong and sometimes hostile reactions from some of the Anglo-Asian community. Carson stresses that Kureishi's presentation of cultural hybridity in his early work is an exploration of the changing nature of both Asian and British identity under the economic crisis and conflict spawned by Thatcherism. In this context *The Buddha of Suburbia's* return to the 1970s seems positive in contrast to the 1980s. With more emphasis on middle class than working class concerns, class is only one of the diverse gender, ethnic and social differences explored in the novel and adaptation. Carson suggests that the mixture of genre interlaces realism with satire and comedy as well as nostalgic elements, and that here this tension between past and present has a postmodern quality in its approach to history. Carson, like Llewellyn-Jones in her chapter on the representation of Ireland, draws upon post-colonial theorists, in particular the connection between hybridities of language and culture significant for Homi K. Bhabha's notion of the 'Third Space' which facilitates the dynamic of constantly changing fluid identities. Similar to the liminal characters in the works described by Tylee, in *The Buddha of Suburbia* it is Karim Amir whose hybrid character and boundary crossings between English and Asian cultures can be read as deconstructive of the notions of 'Otherness' which might be held by some of the audience. In particular, Carson explores the tele-adaptation's representation of Karim Amir's gender and ethnic identity as a tension between a stereotypically exotic 'Other' and the fluidity of his identity. These contradictions most clearly manifest themselves in the processes of male objectification that are linked to comedy as a way of disavowing the homoerotic elements involved in spectator viewing positions.

The extent to which comedy can produce a subversive approach to history is also discussed by Llewellyn-Jones as a part of her comparison of *Ballykissangel* and *Father Ted*, in the post-colonial cultural context from which stereotypical representations of the Irish as 'Other' emerged. The persistence of some of these images is linked to exnomination of post-colonial power relations through the way that TV discourse operates through realist forms. In detailed assessment of different episodes from these successful hybrid comedy programmes, Llewellyn-Jones suggests that where elements of formulaic realism in *Ballykissangel* construct gently comic character identities as commodities to be consumed by the viewer as potential tourist, the exaggerated aspects of critical realism in *Father Ted* subversively draw upon macabre and grotesque elements of the indigenous literary tradition, more in tune with qualities associated with post-colonial theatre practice. Exploration of the relationship of gender stereotypes to the Irish cultural context supports the analysis of these comedies, which includes accounts of their production histories. With brief reference to other programmes and the weighting of their particular generic mixture, the strength of the dynamic between history and humour is discussed as a significant part of the process of the re-invention of Irish identity.

As the relationship between the production companies in Ireland, Northern Ireland and Britain plays a significant role in the distribution and marketing of programmes about Ireland which are shown on different channels, so the process through which American programmes are seen in Britain is crucial to their reception. Janet McCabe's article on the way that 'quality' American programmes are reframed for consumption in

Britain makes important points about how the broadcasting networks shape and manage their schedules by imagining certain audience identities. McCabe suggests that cultural identities are produced, regulated and made visible by institutional discourses. Acknowledging the emergence of cultural identities in Britain in the 1980s, when the competing histories began to be voiced by those groups considered as 'Other', it is argued that it became imperative for the networks to adapt their representations to the needs of their own institutions as well as those of such groups. Through detailed examination of the processes through which *ER* and *The X-Files* were shown on Channel 4 and (originally) BBC2, McCabe's discussion of these programmes indicates how new models of viewer subjectivity can be related to the virtual community, and the ways in which new technology re-defines the relationship between individual and public space. When devoted viewers of *The X-Files* used Internet voting to defeat *Pride and Prejudice* for the 1996 British Academy television awards, McCabe suggests that it made visible the dilemma over the question of quality posed by the conflicting demands of public service and commercial broadcasting.

If segments of the potential are to be increasingly targeted through market research and focus groups on the lines indicated by McCabe, the audience will become even more fragmented in its mirroring of the multiplicity of identities and reading positions. Already it has been reported (*The Guardian*, 17 September 1999) that in his delivery of the Huw Weldon lecture in Cambridge, Mal Young the Head of BBC Drama Series has claimed that in a time of massive change for TV, it is only soap opera which cuts across social boundaries to provide the 'sole remaining shared experience' available to the population, acting as a virtual community. Whether increasing choice and the plethora of postmodern pleasures becoming available through the Internet, interactive TV and the digital revolution will entirely overtake the public service model within the global market place remains to be seen. ITV chief Richard Eyre suggested in his MacTaggart lecture at the *Guardian International Television Festival* in Edinburgh (27 August 1999) that 'Public service broadcasting will soon be dead', pointing out that it relies upon an active broadcaster and a passive viewer. Certainly, as this book suggests, the contemporary viewer is both more knowing and active in selected readings. Eyre's dismissive attitude to regulatory bodies however implies that it is the responsibility of broadcasters to work to a high standard in the public interest- not always an easy stance to maintain in the face of commercial competition. Gavyn Davies' report to the British Government has previously suggested a levy of £24 from all households with digital television, so that this money would subsidise the BBC in an attempt to maintain standards. Jonathan Miller and Nicholas Hellen (*The Observer*, 8 August 1999) considered that Davies' report shows the BBC trails behind all other channels, with even drama only one per cent ahead of the competition in value. They quote Alisdair Milne, a former Director General as saying, 'The BBC... is short of character, imagination and flair'. Their analysis of the situation includes the notion that not only is serious programming in retreat, but the fragmentation of the audience will reduce the opportunities when 'the whole nation can simultaneously cheer the same sporting heroes' or indeed share the same jokes as when Monty Python was a cult programme. It is significant from the perspective of this book that the whole question of quality should thus be linked with maintaining a somewhat spurious and rather old-fashioned sense of national identity, in the face of the emergence of the multiplicity of hybrid identities celebrated in the hybrid forms of contemporary TV drama, which can be more consonant with social and ideological change.

## Notes

1   For example Anthony Giddens, *The Consequence of Modernity*, Cambridge, Polity Press, 1990.
2   See Chapter 6, 'Bardic television', in John Fiske and John Hartley, *Reading Television*, London, Methuen, 1978.
3   Kevin Robins, *Global Times: what in the world's going on?*, in du Gay, P., (ed) *Production of Culture/Cultures of Production*, London, Sage/The Open University, 1997.
4   See Chapter 6, 'Authored drama: not just naturalism', in John Tulloch, *Television Drama: Agency and Change*, London, Routledge, 1990.
5   See Chapter 10, 'Coda – Critical Postmodernism: Critical Realism', in Robin Nelson, *TV Drama in Transition: Forms, Values and Cultural Change*, London, Macmillan,1997.

## Bibliography

Alvarado, M., & Stewart, J. (eds), *Made for Television: Euston Films Limited*, London, British Film Institute, 1985.

Brunsdon, C., 'Problems with Quality', *Screen* 30/1, 1990, pp 67-90.

Caughie, J., 'The Logic of Convergence' in Hill, J. and McLoone, M., (eds), *Big Picture, Small Screen: The Relations between Film and Television*, Luton, John Libbey/University of Luton Press, 1996, pp 215–223.

Caughie, J., 'Before the Golden Age: Early Television Drama', in Corner, J., (ed), *Popular Television in Britain: Studies in Cultural History*, London, British Film Institute, 1991, pp 22–41.

Fiske, J., and Hartley, J., *Reading Television*, London, Methuen, 1978.

Geraghty, C., *Women and Soap Opera*, Cambridge, Polity Press, 1991.

David Graham and Associates, *Building a global audience: British television in overseas markets*, Department for Culture, Media and Sport, Broadcasting Policy Division, 1999.

Hebdige, D., 'Towards a Cartography of Taste 1935 – 62', in Hebdige, D., (ed), *Hiding in the Light*, London, Comedia/Routledge, 1988.

Hill, J., 'British Television and Film: The Making of a Relationship', in Hill, J. and McLoone, M. (eds), op. cit., pp 151–176.

Kerr, P., 'Classic Serials – To Be Continued', *Screen* 23/1, 1982, pp 6–19.

Leonard, M., *Britain™*, Demos, 1997.

# Fact, Fiction and

# the Ideology of Identity

## Docudrama as Melodrama: Representing Princess Diana and Margaret Thatcher

Jonathan Bignell

### Introduction

This chapter analyses the blending of documentary re-construction and melodramatic form in television representations of Princess Diana and Margaret Thatcher, focusing on *Thatcher: The Final Days* (1991) and *Diana: Her True Story* (1993). The dramas were promoted as 'factual' documents of the personal lives of the two familiar but inaccessible women. However, the form of both dramas is that of melodrama, a fiction genre marked by its focus on women characters, its focus on the emotional and the psychological, and its emphasis on moments of dramatic intensity. Thus there are divergent genre forms in the dramas, paralleling the divergence between the discourses of current affairs and of popular personality reportage, the two main media forms through which these women have been represented. The chapter argues that this mixing exposes the contradictions inherent in contemporary culture's representations of iconic female figures.

As Derek Paget (1998: 61) has described, the aim of drama-documentary on television has been to 're-tell events from national or international histories' and/or 'to re-present the careers of significant national or international figures' in order to review or celebrate these people and events. The key figures and important moments depicted are often familiar to the audience, and close in time to the transmission of the programme. Devices like opening statements and captions make clear the factual basis of docudramas, while disclaimers state that some events have been changed or telescoped, and some characters may be amalgamations or inventions. Paget's definition of drama-documentary is that it 'uses the sequence of events from a real historical occurrence or situation and the identities of the protagonists to underpin a film script intended to provoke debate.... The resultant film usually follows a cinematic narrative structure and employs the standard naturalist/realist performance techniques of screen drama' (p.82). While these outlines go a long way to describe the ways that drama-documentary works, a significant shift in dramatic mode occurred when the subjects of docudrama were Princess Diana and Margaret Thatcher. The dominant mode of the programme became melodrama, rather than the usual naturalism of film and television fiction. This essay analyses the blending of 'factual' reconstruction and melodramatic form in television representations of two iconic women, Princess Diana and Margaret Thatcher, focusing on *Thatcher: The Final Days* (Granada, 1991) and *Diana: Her True Story* (Sky 1, 1993).

The *Thatcher* drama was based on documentary records and interview information, and its opening on-screen caption describes the drama as a 'dramatic reconstruction of three weeks in Autumn 1990 which led to the downfall of Prime Minister Margaret Thatcher'. *Diana: Her True Story* was based on Andrew Morton's bestselling book of the same title which drew on interviews with Diana and her circle, and dramatised her life from the separation of her mother and father during her childhood up to her separation from Prince Charles. The dramas were promoted as factual documents of the personal struggles behind the scenes, in the lives of two familiar and public, but inaccessible, women. All roles were played by actors in each drama, though for clarity I will refer here to characters' names throughout. The mode of both dramas has much in common with melodrama, which in television is marked by its focus on women characters, on the emotional and the psychological, and on moments of dramatic intensity. *Thatcher* and *Diana* drew on the two dominant media forms through which these women have been represented. Their documentary base is signalled by opening statements about the accuracy of the stories, and by the appearance of journalists and TV cameras within the dramas, where the world of the news media is always ready to intrude upon and comment on the actions of the central figures. The melodramatic mode, on the other hand, draws on popular gossip and personality reportage about the personal lives and characters of Margaret Thatcher and Princess Diana, supplying the means to interpret dramatic turning-points and crises through a repertoire of stock character-types and familiar codes of gesture and expression.

The reason that the modes of fact-based docudrama and melodrama appear together in the two programmes seems to be that they are based on women's lives. Margaret Thatcher and Princess Diana, despite many differences between them, were both in the unusual position of being very publicly visible in a conventionally masculine environment. Media attention focused on their private lives as wives and mothers, as well as on their public roles. So *Thatcher* and *Diana* offer the pleasure of recognising familiar figures, events and issues in the public realms of politics and elite institutions, and also the pleasures of identification and fantasy focused through the private experience of these public women. The history of television features on the two women shows an interest in both their public roles and their private lives and personalities, a combination almost unthinkable with male political and constitutional figures. There have been many programmes on Diana both before and after the *Diana* dramadoc, including *Diana: The Making of a Princess* (1989), *Diana: Progress of a Princess* (1991), *Diana: Portrait of a Princess* (1994), and many tribute programmes after her death, including *Diana: A Celebration* (1997). The very titles of many of these programmes link the personal (the first name) with the public (the royal title). Similarly, Margaret Thatcher has featured as the subject of documentary and personality-profile programmes including *The Thatcher Factor*

(1989), *Thatcher: The Downing Street Years* (1993), and *Thatcher: The Path to Power – and Beyond* (1995). Political issues were personified, and some programmes took a special interest in Margaret Thatcher's use of 'housewife and mother' roles in both her public and her private self-presentation. There is nothing of Thatcher's private life in *Thatcher: The Final Days*, but instead the political environment is depicted in the familial and domestic terms of US prime-time melodramas like *Dallas* or *Dynasty*. Discourses of femininity were important to the public images of both Diana and Margaret Thatcher, and the dramadocs follow the linkage between public and private in their linkage of documentary and melodrama.

Television melodrama is continuing drama, but *Thatcher* was a single programme broadcast on the ITV network and not a continuing serial, while *Diana* was originally a mini-series for Sky Television, now sold as a video film, so there seems little reason to parallel these docudramas with prime-time melodrama serials at the level of narrative form. However *Thatcher* and *Diana* each show a part of a longer-running story. Margaret Thatcher's rise to the leadership of the Conservative party, her time as Prime Minister, and her subsequent fall; Diana's relationship with Prince Charles, her marriage, and her departure from the royal family. Unlike *Dallas* or *Dynasty*, docudramas are not assumed to address a mainly female audience, nor do they foreground relationships between women and among extended families. But as in US prime-time melodramas like *Dynasty*, *Thatcher* and *Diana* feature powerful female protagonists in elite environments (like royal palaces and the Houses of Parliament). In a review of *Thatcher*, Lynne Truss remarked that 'the programme's most lasting impression may derive from the wallpaper and Regency chairs' (1991: 15), and great efforts were made in *Diana* to use sets which reproduced the interiors of famous stately homes. Andrew Morton's introductory piece to camera in *Diana* reports that 'no expense has been spared' in an effort to present 'an idea of the richness of Royal life but also the human drama'. Nevertheless, parallels between television melodrama and these docudramas are most clear in relation to the portrayal of the central women characters.

As Ien Ang's (1997) research on *Dallas* showed, the appeal of prime-time melodrama rests primarily on the central women characters, particularly Sue Ellen, the wife of the villainous J. R. Ewing. The viewers who communicated with Ang 'assert that the appeal of Sue Ellen is related to a form of realism (in the sense of psychological believability and recognizability); more importantly, this realism is connected with a somewhat tragic reading of Sue Ellen's life, emphasizing her problems and troubles' (p.157). Morton's introduction to *Diana* alludes to this narrative mode by describing the drama as 'a vivid human interest story about a dream marriage that turned into hell for Diana: it's a story of a fractured fairytale'. Sue Ellen hated her husband but lacked the strength to leave him. She was a successful businesswoman in a masculine world, but her business (Valentine Lingerie) was set up as a tactic to separate J. R. from her rival, his mistress. While activity in the masculine world of business intrigue was present, its significance lay in its relationship to Sue Ellen's personal dilemmas. Crude parallels could be drawn here with Princess Diana's increasing stature as an independent player in campaigns and charitable work in the public eye, activities which were seen by the media as in part attacks on Diana's husband, Prince Charles, who was having an affair with Camilla

Parker-Bowles. Just as Sue Ellen became an alcoholic in response to her relationship problems, it is claimed that Princess Diana became bulimic and occasionally suicidal. In *Diana*, we see, for example, a sequence of scenes aboard the Royal Yacht where Diana first discovers cufflinks on Prince Charles's shirt featuring the intertwined Cs of Charles and Camilla. 'You pig!' she screams, and walks out of the room. The scene cuts to Diana voraciously eating cake in the Yacht's kitchen, accompanied by foreboding music in a minor key. The next shot is of Diana leaving a toilet, whose flush is heard in the background. Cause and effect are established rapidly by the sequence, attributing Diana's physical problems to emotional disturbances provoked by her husband. The privileging of Diana's emotional life in the drama is especially apparent in relation to the 'love-triangle' of Diana, Charles, and Camilla, so that when Diana hears Charles on the telephone to Camilla saying he will always love her, Diana's response is to confront him saying 'What about our life together? I need you to look after me, I need you to hold me, I need you to touch me', Charles replies 'You're always being sick', to which Diana responds 'I feel so abandoned'. As in prime-time melodrama, conflict between characters produces emotional drama in *Diana*, and characters also experience conflicts within themselves which are expressed by rapidly alternating and conflicting emotions, often expressed through physical, bodily behaviour.

In the other major prime-time melodrama of the 1980s, *Dynasty*, the central woman character was Alexis Colby, who the audience was invited to both admire and despise. Jostein Gripsrud (1995: 231), quoting two other writers on soap melodrama, comments on Alexis's ability 'to "transform traditional feminine weaknesses into the sources of her strength" (Modleski, 1982: 95), i.e. her uninhibited use of her (and men's) sexuality in her struggle for power, is combined with "her own skills at the kind of business manoeuvring which was previously deemed to be a masculine prerogative" (Geraghty, 1991: 136)'. Alexis was regarded as the ultimate bitch, aggressive in a masculine manner, sexually manipulative, but all because of her untimely separation from her beloved children. In other words, her masculine behaviour was the result of a thwarted and distorted femininity. Similarly, Margaret Thatcher was represented as a masculinised woman, reputed to be domineering and ruthless, to the extent that satirists sometimes portrayed her in men's clothing or with a male body. Marina Warner (1987: 51-2) commented:

> It may seem paradoxical to argue that Mrs Thatcher's femininity matters, since it is more usual to note her determination, toughness, dynamism and strength, and then to assert that this is characteristic masculine equipment, making her female by sex alone. ... It is she who is 'the best man in Britain'; she who 'wears the trousers'. But the conundrum she poses is more complex: she never wears the trousers. She is careful to live up to the conventional image of good behaviour in women prescribed for Conservative supporters.

Representations of Margaret Thatcher, including during *Thatcher*, focus on the same apparent contradiction which makes Alexis in *Dynasty* so fascinating, the contrast between masculinity as a role, and the femininity expected of women. Just as Alexis was determined to control the Colbyco corporation, and manipulated the men around her with ruthless skill, Margaret Thatcher maintained control of the male-dominated and patriarchal Conservative party and the Government. At a Cabinet meeting near the beginning of the drama, Thatcher complains that the drafting of bills is behindhand, asking sharply 'Would someone care to tell me why?' A series of brief medium shots follows, showing the assembled Ministers looking down at their notes sheepishly, or fiddling with their papers to avoid her gaze. While the scene is based on factual evidence, its dramatic significance comes from the characterisation of Thatcher as a domineering boss in the manner of Alexis Colby. Thatcher and Diana are presented in the terms of mythic kinds of femininity which the viewer can recognise from television melodrama and other media sources.

The melodramatic characterisations in *Thatcher* and *Diana* explore public roles by grounding them in personal characteristics, expressed through declarations of will, reactions to dramatic reversals, and by emphasising gestural signification, the dominant expressive form in melodrama. In a dramatisation of an interview with a *Times* journalist, Thatcher's familiar patterns of speech and gesture, and her political dogmatism, are brought together in her reaction to Michael Hestletine's candidacy for Conservative Party leadership. She leans forward, shot in close-up, speaking loudly and emphatically, saying 'We cannot go that way, we *cannot go that way*', then breaks into a confident smile. Diana's characteristic glance from under the fringe of her hair, and her relative awkwardness in her youth versus a more confident bearing later in life, are used in *Diana* both to recall media images of her on television and in the press, and to chart her emotional development. Since media icons like politicians, film stars, and members of the royal family are recognised by their characteristic media images, their representations are already composed of a restricted repertoire of facial expressions, tones of voice, and gestures, like the repertoire of characteristics which define melodrama characters. *Thatcher* shows Mrs Thatcher in a simulated broadcast of a parliamentary debate where she famously said 'No, no, no' to European integration, where the camera angle and shot type exactly match the conventions of the real footage. Re-enacted moments in *Diana* include the positioning of the camera to duplicate the famous press photograph showing her legs through a see-through skirt, to the Royal Wedding itself where close-ups on the actors portraying Diana and her father are carefully integrated with broadcast coverage of the event. Reference to the images that the audience already know, together with the actors' mimicking of familiar bodily and facial movements, and tones of voice, both aids documentary realism and triggers the audience's response to the central figures in terms of the restricted but powerful language of television melodrama.

*Thatcher* was made by journalists from Granada Television's *World in Action* current affairs documentary series, in collaboration with Granada's drama department, for a budget of £500,000. The mixture of journalistic and dramatic modes led Lynne Truss, reviewing *Thatcher*, to describe the experience as like 'watching [the political satire programme] *Spitting Image* with the sound turned off', and 'looking forward to (and checking off) familiar scenes' from television news footage. *Thatcher* comes from a tradition of Granada Television fact-based dramas, which derive their authority from Granada's current affairs programming. The central figures in this tradition conceive docudrama to be based in immaculately researched journalistic investigation (Paget 1998: 165-8). *Thatcher* is a drama based on real events, and legitimates itself with the apparatus of television journalism including an emphasis on exactitude of chronology and the sequential unfolding of events. Date captions are very common at the beginnings of scenes, and captions also identify the names and job titles of politicians and civil servants. But Richard Maher, the scriptwriter of *Thatcher*, described the story as 'a tragedy of hubris' (Anon., 1991: 15), referring explicitly to dramatic tradition. The tragedy was caused, he felt, by the fatal flaw of Thatcher's pride, which led her to allow her parliamentary private secretary Sir Peter Morrison to reject help from her media advisors Sir Tim Bell and Sir Gordon Reece, who had masterminded her election victories. A key moment is the scene when the camera is positioned among the television news crews gathered to witness Michael Hestletine announcing his candidature in the leadership battle. This moment, where the drama involves the audience among news media personnel jostling for the best position, is juxtaposed with the following scene in which Peter Morrison is on the phone to Tim Bell, and tells him 'This really isn't your line of country at all.... It isn't a media battle', though the audience has just been shown that it is precisely a media battle. The isolation of Thatcher from the public relations battlefield, from the electorate, and from her fellow Members of Parliament, is demonstrated by dramatic juxtaposition and cutting between thematically related scenes. Despite this emphasis on conventional dramatic structure, Maher emphasised the factual basis of the script: 'We have done our absolute best to make sure that everything you see actually happened, and the main reason anyone would want to watch this programme is that it's true, even at the risk of being boring' (Anon., 1991: 15). Apart from Thatcher, almost every figure represented in the programme was involved in its making, contributing diary notes and memories to the script. But dramatic crises are integral to the programme, and include the spectacle of John Gummer bursting into tears in his interview with Thatcher, part of a series of meetings where members of the Cabinet express their support for her, or lack of it. Thatcher asks Kenneth Clarke, 'Who says I should go? *Who says I should?*', establishing a pathetic distance between her own view of her position as unassailable and the desertion of her by her erstwhile colleagues.

*Diana: Her True Story* is more closely related to US television docudramas about sensational news events. Producers of these dramadocs work differently to the British Granada tradition: 'Protecting themselves legally by using real-world protagonists as consultants at the point of production, they buy the rights to a point of view rather than subscribing to the older journalistic notion of "objective" accounts of events in the news' (Paget, 1998: 196). While *Thatcher* showed no events which could not be confirmed by two or more sources, *Diana* relied heavily on Diana's own point of view, and used few written sources. Andrew Morton had worked for a decade as a 'royal reporter' before publishing the book version of *Diana: Her True Story*, selling stories to a range of British newspapers, and was not an investigative journalist in the mould of the *World in Action* team. Kevin Connor, the director of *Diana*, had previously worked on literary adaptations like *Great Expectations* (1989), was experienced in high-budget docudrama based on recent events like *Iran: Days of Crisis* (1991), and had made the mini-series melodrama *North and South – Book 2* (1986). The video release of *Diana: Her True Story* shares the same cover photograph as Morton's book on which it is based, and has a novelistic structure, beginning with Diana's childhood fears of abandonment, documenting her unhappy marriage, and concluding with her eventual independence. Morton's introduction to the drama describes it as the 'story of a girl who became a princess before she became a woman, and a woman who found herself, in the face of adversity'. Diana's last line, spoken as she leaves a meeting with the Queen, Prince Philip and Prince Charles about the terms of the couple's separation, is 'When I turn off my light at night, I'd like to know that I've done my best'. The story of personal growth and self-realisation relies on an ideology of expressive individualism which it shares with the nineteenth-century novel, contemporary romance fiction, and American television melodramas and mini-series. The representation of Princess Diana in *Diana* has more in common with television fiction and popular journalism than with the current affairs tradition from which *Thatcher* derives. In effect, *Diana* is Diana's version of her own life, a part of the press and public relations battle between her and the royal family. Diana's life had been massively covered by the media for many years before *Diana* was made (see Franklin, 1997: 218-222; McNair, 1998). Both Thatcher and Diana are portrayed in the context of a contest of representations around them, and in *Diana* a representative photo-journalist character is incorporated into the drama to comment on her evolving relationship with Prince Charles, remarking finally in 1987 that 'All that's left of this fairytale marriage is a photo-opportunity'. While the media are not represented as forces working for Thatcher's or Diana's downfall, the media's ubiquitous presence to narrate and comment on the action in *Thatcher* and *Diana* acts as a kind of chorus, a public voice like the choruses of citizens in Greek tragedy, making the link between the personal dramas of an elite and their relationship to public political life.

In Britain there is a long-standing view that public life should be taken seriously by the media, and there is a temptation to criticise the melodramatic mode in *Diana* and *Thatcher* on the assumption that fact-based drama should be journalistically objective as against the perceived triviality and partiality of American dramadoc. But this view also reveals an implicit bias against the perceived femininity and low status of melodrama as a dramatic mode. The expression of emotion, as David Lusted (1998) notes, is both a marker of femininity and of working-class culture. While masculine values (like those of politics, journalism, and the British royal circle) entail the suppression of emotion in favour of efficiency, achievement, and stoicism, feminine values encourage the display of emotion as a way of responding to problematic situations. Similarly, elite class sectors value rational talk and writing as means of expression, versus emotional release. These distinctions, which are of course culturally produced rather than biological or natural, have been important to work in television studies on the relationship between gender and the different genres of television, where news and current affairs are regarded as masculine, and melodrama as feminine. On the basis of these gender, class, and genre distinctions, the role of emotional display in *Thatcher* and *Diana* takes on increased significance. Diana's frequent tearful outbursts (and her struggles against them) separate her from the stoical elite group which surrounds her, some of whom are also women, and parallel her with the ordinary viewer. Thatcher's eventual capitulation to tears at the final meeting with her Cabinet is also a marker of her defeat by masculine forces and the values of the political culture which she had sought to control. She is seen in medium shot across the cabinet table, making a final statement before withdrawing from the leadership contest, remarking, when her voice breaks, 'I've never done that before'. Christine Geraghty (1991: 74) argues of US prime-time soaps that they are set in a world controlled predominantly by men, but offer pleasures to the woman viewer by showing that male power can be challenged 'on the one hand by moral questioning and on the other by women's refusal to be controlled'. In contrast to this, *Thatcher* and *Diana* show women failing to hold onto their power within masculine elites, and accepting more or less dignified ways out from them. The dramatic climaxes of *Thatcher* and *Diana* attain their climactic status at the cost of the ejection of the women from the masculine public world.

Alexis in *Dynasty* can be read as a tongue-in-cheek critique of patriarchy in her use of feminine and masculine gender traits as weapons against male competitors, where her vampish otherness succeeds in getting her what she wants. This representation of Mrs Thatcher is also to be found in *Thatcher*, but her emasculation of male competitors still leaves some who are able to expose her apparent narcissism and self-delusion, marked as feminine characteristics, and thus defeat her. *Thatcher* features a group of male Conservative politicians who are competing and plotting in a similar way to the corporate battles in 1980s television melodrama. In *Thatcher* the viewer sees Geoffrey Howe watching a recreation of Thatcher's televised Guildhall speech of 12 November, in which she says 'I'm still at the crease, though the bowling has been a little hostile of late. And in case anyone doubted it, I can assure you there will be no ducking the bouncers, no stonewalling, no playing for time. The bowling's going to be hit all round the ground. That's my style'. As the laughter of her audience dies away, the scene cuts to Howe's own resignation speech the following day in the House of Commons, which picks up her cricketing metaphor and uses it to critique her leadership style: 'It's rather like sending your opening batsmen to the crease, only to find, the moment the first balls are bowled, that their bats have been broken before the game by the team captain'. In *Dynasty* and *Dallas*, the family is the site of both economic struggle and moral corruption, and the narrative of the serial depends not on the integration but the disintegration of the family. *Thatcher* tells the story of the disintegration of the matriarch's authority over the Conservative party, while powerful male figures adopt different masculine roles to submit to or challenge her authority. Hestletine is portrayed as a ruthless youngster, Howe a weak and vindictive father, while Gummer is in love with Mrs Thatcher's phallic mother figure. John Major becomes the quiet favourite son complete with vyella pyjamas. Like the fight for Ewing Oil, the fight for Conservative party leadership is strongly gender-inflected, and draws in part on a parallel between the ruling elite and an extended family. In the commercial break between parts one and two of *Thatcher*, the audience was coincidentally reminded of *Dallas* by the appearance of Larry Hagman, in a role recalling his performance as J. R. Ewing, in an advertisement for British Gas heating.

At the height of Diana's popularity, Ros Coward (1984: 163) paralleled the royal family with prime-time soap opera: 'The two soap operas share the same preoccupations: the unity of the family; family wealth; dynastic considerations like inheritance and fertility; sexual promiscuity; family duty; and alliances with outsiders/rivals/lower orders. The fact that "The Royals" is loosely based on reality only adds to its fascination.'

## Docudrama as Melodrama: Representing Princess Diana and Margaret Thatcher

Whereas the Ewing, Carrington, and Colby families in *Dallas* and *Dynasty* are elevated by extreme wealth, the royal family are elevated by nobility, but each family drama is about how ordinary conflicts are lived out when the family seems to have everything it could want. *Diana* shows her on the 1983 Australian tour accepting a bouquet from a little girl who was desperate to meet her, where Diana tells her 'I know what it's like to really want something' , to which the girl replies 'But you have everything'. Like *Dallas*, *Diana* works with the paradox that the royal family are immensely privileged and different from the television audience, yet they appear to share the emotional, familial and professional pressures experienced by ordinary people. In working through these problems in narrative, melodrama presents characters as simplified archetypes, for example as good mother or bad mother, faithful spouse or unfaithful spouse, conformist or rebel, princess or bitch. *Diana* establishes Prince Charles first as an attractive and eligible bachelor, then after his marriage as a domineering unfaithful husband. Diana is represented first as a virginal ingenue, then as a scold or a victim during her marriage, and finally as a modern and liberated heroine.

Christine Geraghty (1991) and other feminist critics have argued that the world depicted in melodrama is potentially Utopian, since the suggestion is always there that it could be reorganised in terms sympathetic to women's desires for community, and for openness and honesty of feeling. But neither *Thatcher* nor *Diana* suggest this. In *Thatcher* and *Diana*, docudrama is fractured by the personalising discourse of melodrama, and public figures are explained by their personal psychologies. Individuals and familial structures, riven by the conflicts which are the stuff of television melodrama, become the ways of making sense of the ruling Conservative party and the royal family. The audience sees the public worlds of the Conservative government and the royal family through the dominant modes of the different TV genres of documentary reconstruction and melodrama, so that any single understanding of the world might be revealed as flawed. But there is no suggestion in either docudrama of a politics which would address this decoupling of reality and representation, or of a politics of television form (Caughie 1981) which could bridge the gulf between the gendered modes of docudrama and melodrama. This was not the case with the media representations of Princess Diana's death, for example, which enabled not only conservative support of institutions like the royal family or the political establishment, but also a degree of challenge to them, reversing binary structures of power like male/female, Crown/subjects, public/private, or rational/emotional. Seeking framing narratives for the event, the media explained the public reaction to Diana's death in the terms of melodrama, as the sign of a frustration with depersonalised and detached culture, leading to a desire for engagement, feeling and passion, and a longing for collectivity and unity (Becker, 1998; Hermes, 1998). It was in live television coverage and news and current affairs programming that the media were forced to confront the erosion of differences between documentary objectivity on the one hand, and on the other hand melodramatic excess and sentiment, which had already been signalled in *Thatcher: The Final Days* and *Diana: Her True Story*.

## Bibliography

Anon., 'Dramatic Licence', *The Times*, 11 September 1991, p15.

Ang, I., 'Melodramatic Identifications: Television Fiction and Women's Fantasy' in Brunsdon, C., D'Acci, J., & Spigel, L., (eds) *Feminist Television Criticism: A Reader*, Oxford, Oxford University Press, 1997, pp155-66.

Becker, K., 'Ritual', *Screen* 39:3, 1998, pp289-93.

Caughie, J., 'Progressive Television and Documentary Drama' in Bennett, T., Boyd-Bowman, S., Mercer, C., & Woollacott, J. (eds) *Popular Television and Film*, London, BFI Publishing & Open University Press, 1981, pp327-52.

Coward, R., 'The Royals' in Coward, R., *Female Desire: Women's Sexuality Today*, London, Paladin, 1984, pp163-71.

Franklin, B., *Newszak and News Media*, London, Arnold, 1997.

Geraghty, C., *Women and Soap Opera: A Study of Prime Time Soaps*, Cambridge, Polity Press, 1991.

Gripsrud, J., *The Dynasty Years: Hollywood Television and Critical Media Studies*, London, Routledge, 1995.

Hermes, J., 'Hollywood', *Screen* 39:3,1998, pp293-5.

Lusted, D., 'The Popular Culture Debate and Light Entertainment on Television' in Geraghty, C. & Lusted, D. (eds) *The Television Studies Book* London, Arnold, 1998, pp175-90.

McNair, B., *The Sociology of Journalism*, London, Arnold, 1998.

Modleski, T., *Loving with a Vengeance*, Hamden, Conn., Shoe String Press, 1982.

Paget, D., *No Other Way to Tell It: Dramadoc/Docudrama on Television*, Manchester, Manchester University Press, 1998.

Truss, L., 'Mute, Inglorious', *The Times*, 12 September 1991, p15.

Warner, M., *Monuments and Maidens: The Allegory of the Female Form*, London, Picador, 1987.

# Docusoap: Actuality and the Serial Format

## Gail Coles

### Introduction

Docusoap, as the name implies, is an example of the current hybridisation of television genres which, while worrying for some, can also be seen as a revitalisation of the pleasures of both documentary and drama for audiences. This chapter examines how docusoaps span the drama/documentary divide to portray the everyday experiences of work, family and relationships in the lives of their 'real' subjects. The actuality at the heart of docusoaps, expressed through their observational film-making techniques, is combined with the narrative structure of soap opera, and focused through the extraordinary 'ordinary people' at the centre of the programmes. It is this combination of elements that accounts for their wide popular appeal and which transforms them into ratings phenomena. It is a small step from the fictionalisation of soap narratives to the 'living soaps' of today's TV schedules.

> A lot of these shows set out to generate reactions rather than thought, favour emotions over ideas. Go for moments, rather than story. The worst ones aim to give the audience its jollies in cheap thrills and regard the ordinary people they briefly focus on purely as high yield, low-cost minutage. And what else do they have in common? Oh yeah. They get good figures. That's where my argument breaks down (Hamilton, 1998).

Docusoap has been widely attacked since it emerged as a new television format just three years ago. Critics, such as comedy writer Andy Hamilton (quoted above), use docusoap as an example of the 'dumbing down' of British broadcasting. They blame an increase in cheap, exploitative entertainment programming on the new competitive and deregulated climate. Docusoaps are, as Hamilton also points out, extremely popular with audiences. Programmes such as *Driving School* (BBC 1, 1997), *Airport* (BBC 1, first series 1996) and *Vets School* (BBC 1, 1996) have been extremely successful in the ratings. This popularity is often explained either by reference to the audience's stupidity or to their baseness. Popularity in itself cannot be a measure of innovation or value, but a more critical examination of the appeal of the format seems a useful first step in understanding the contribution these programmes make to television and its audience.

The proliferation of docusoaps came out of a crisis in early evening programming. Sitcoms and game shows, the usual pre-watershed fare, were failing to get substantial audiences. By combining actuality with elements of comedy and spectacle, in a serial format, early evening schedules were invigorated. But what is it about this formula that appeals to audiences? What does this popularity reveal about audiences' changing tastes, and their expectations of, and uses for, television programming? What does their prolif-eration through the schedules tell us about how broadcasters perceive their audience and cater for it, within present economic constraints? The newness of docusoap means little has been said about it beyond comments by journalists and programme-makers, made either to condemn programmes or defend them. My aim is to begin to map out the terrain for a more considered discussion, taking into account the difficulties of determining the contours of such a recent genre, one that is still developing and changing.

## 1990s Television

The term docusoap may be a journalistic invention, but it provides a useful insight into the formula and its appeal. This is a format combining two of the most popular television genres, reality programming and soap opera, into a successful hybrid form. This hybridi-sation is an important characteristic of not only docusoap, but other television genres, as well as cultural production generally in the 1990s. It is a defining aspect of what is called our post-modern age, in which the boundaries between texts, as traditionally formulated by critics and programme-makers and generally understood by their consumers, have become increasingly blurred, with the inclusion of styles and subjects from other texts and genres. Allied to this 'intertextuality' is an awareness in the text of its own processes and sources, which are read and appreciated by audiences. Whereas hybridisation as part of generic development and change is not new, one can claim that this reflexivity is.

Much of the concern about docusoap is the result of its categorisation as a debased form of documentary, which limits a more constructive examination of the forms, subjects, intentions and consumption of the programmes. Approaching docusoap as a new hybrid genre incorporating the actuality of documentary and the melo(drama) of soap opera, lightened by the humour of sitcom and performances associated with light entertainment, will provide answers to the questions about television and its relationship to its audience(s) raised earlier.

It is important to note that as well as the merging of formats in television programmes, there have been changes in the relationship between texts and viewers which have had an impact on their form. Viewers are now able to schedule programmes at their own conve-nience through the use of VCRs, as well as 'surf' through programmes on an increasing number of terrestrial, satellite, cable and digital channels. With the migration of sets from the family sitting room to other rooms in the house and the increasing convergence of television with telephony and computers, television viewing has become a more frag-mented, less self-contained experience. Programme-makers have incorporated this more distracted relationship between viewer and programme into the aesthetics of the programmes. Using short segments, fast-paced editing and more fluid camera move-ments, the programmes produce a visually stimulating style which can work to arrest viewers moving through the channels. The short, discrete segments also allow viewers to enter into the programmes and to make sense of them at various points, rather than demanding that they begin watching at the beginning.

These changes in television viewing have their roots in new media technologies and changes in lifestyle, in which television functions as just one of a number of leisure pursuits. The mass or family audience has fragmented into diverse, demographic groups according to age, gender, sexuality, ethnic or regional identity. Whether or not the actual audience has changed, or is merely perceived to have done so by broadcasters, and therefore addressed differently by them, is an open question (Feuer, 1992:153-154). But what is clear is that the increased visibility and vocalness of groups organised around issues of cultural identity, combined with the destabilisation of the establishment which had dominated British television, has resulted in a loss of confidence amongst broadcasters about what their audience wants and needs. This has resulted in an increased emphasis on ratings as a way of determining viewer interests and maintaining a large audience share in the current competitive television climate.

Contemporary television viewers also comprise a more tele-literate audience. They are familiar with the history and conventions of television, and enjoy seeing them referenced in new and unusual ways. The most vibrant television forms in the last ten years have been fly-on-the-wall observational programmes and soap opera, both of which contribute to the docusoap. These formats are relatively new and strongly associated with the medium of television. As such they are more adaptable than other dramatic or entertainment forms which have longer histories and stronger traditions. Documentary, in its role of representing and articulating social and political events and issues, has always had to be able to reinvent itself to adapt to its time. As a cheaper form than drama, it has often been an entry point for new film- and programme-makers, which also contributes to its reputation for innovation. Soap opera is also a relative newcomer and its low status as a popular dramatic form means it is less restricted by notions of taste and propriety. Although soap opera is highly formularised, programmes have recently incorporated difficult and emotive issues into their narratives. Thus the coming together of actuality and dramatic formats results in a genre which is both familiar and innovative to audiences. The actuality of the observational documentary, the narrative and aesthetic elements of soap opera, the situation and humour of situation comedy, combined with the performance of a found 'star' such as Maureen (*Driving School*) or Jane (*The Cruise*), turn these programmes from the ordinary to the phenomenal in terms of ratings.

## Reality/Actuality

Richard Kilborn and John Izod, discussing the audience reception of documentary, utilise the concept of 'frame and context' (1997:32/33) to understand the practice of decoding media communication. The frame refers to the 'mental framework' within which viewers approach texts. This frame is produced through textual codes, and by viewers' own histories through which they read the codes. The context refers to the extratextual information available through advertising, and word of mouth, as well as the actual viewing context in which the programmes are consumed. Thus the 1990s world of television with its hybrid forms and distracted patterns of viewing, has an impact on the frame and context which shape audience reception. Equally important to their frame and context is the fact that docusoaps are commissioned by documentary departments, are produced and directed by documentary film-makers, utilise the current technology and production techniques of *verité* documentary, digital cameras and small crews, and are advertised both within television and in other media as documentary.

Docusoaps incorporate the formal characteristics of documentary in their use of actual locations and 'ordinary' people, and utilise an observational style associated with a relatively unmediated representation of those subjects. Like documentaries, docusoaps claim to tell us something truthful about the world we live in and the people who inhabit it. 'At best they entertain, but they can also examine the interaction between us all as individuals and in the workplace' (Jeremy Mills, Executive Producer of *Hotel*, quoted in Dams, 1998). Drama often makes a similar claim, but documentary promises to capture that truth through actuality rather than through fictionalised scenarios.

Whereas documentary tends to use subjects as representing a position in an argument (Kilborn & Izod, 1997: Chapter 5), docusoaps privilege character(isation) and personal relationships over social or political issues. Thus in much documentary, participants' backgrounds and personalities are only relevant in so far as they contribute to the argument being pursued; in docusoap, the participants and their daily lives are the focus. Personal issues such as love and marriage feature as a large part of the hermeneutics of these programmes. Will the young *Holiday Reps* (BBC1, 1997) find true love and lasting relationships with their local boyfriends, in spite of all we know about the vagaries of holiday romances? These questions are raised, explored, and delayed in much the same way as they are in serial dramas. The issues cross over episodes, ending in cliffhangers, until some resolutions are provided, in this case in the special edition, *Holiday Reps Get Married* (BBC 1, 24th March, 1999). Although like soap opera resolutions, partnerships (Debby and Marcos), new careers (Caroline) and even weddings (Eve and Andreas) remain contingent and uncertain.

Television has always been focused on the domestic sphere, but has, in the last two decades, taken an increased interest in the personal. New technology has assisted in this pursuit. Allied to this is the change in public opinion about what is acceptable, tasteful, and appropriate for public viewing. Just as the technology available to the British Documentary Movement film-makers of the 1930s and 1940s affected their realist style, and the new lightweight technology of the 1950s and 1960s underwrote the ethos of *cinéma verité* and direct cinema, the 1990s world of docusoaps has its own technology. The new lightweight DV cameras, the possibility of smaller crews or even no crews at all (CCTV or subjects filming themselves in video diaries) means television can record events in spaces previously impossible.

Jean Baudrillard, writing in the early 1980s, used the American fly-on-the wall series, *An American Family* (PBS, 1973), to describe not only this '...*perverse* pleasure of spying' (1983:50), but the collapse of boundaries, discussed earlier in this chapter, that characterises much of 1990s television. The boundaries to which Baudrillard refers are those between life lived on camera and life lived off camera, and ultimately to the blurring of boundaries between the real and its simulation or pretense. This merging of the real and its TV performance is directly related to the technology and ethos of Direct Cinema or fly-on-the-wall as it has come to be known. This is observational cinema which pretends that although we are there, the camera and crew are not, and therefore are not affecting events. This pretence is maintained by refusing any direct address to the camera.

Although docusoaps have a direct link to fly-on-the-wall film-making techniques, through their look of action caught unawares, they have combined that with more reflexive techniques which make the film-making process visible. This is emphasised through direct address by subjects to the director or crew, or in video diary formats, directly to the viewer. Chris Terrill, docusoap film-maker, has referred to the camera as 'another charac-ter' in the production rather than a silent witness (*On Air*, BBC 2, 1998, 'The Truth Behind TV: Docusoaps'). In addition, the excessive performance styles of many subjects of docu-soap, combined with the media awareness of the contemporary audience, produces a self-consciousness and ironic presentation which reintroduces the gap between the real and its simulation for both participants and viewers.

As well as referencing traditional documentary styles, docusoaps are part of the expanding area called popular factual programming, which combines information and entertainment to reach a wider audience, often disparagingly termed 'infotainment'. They have important links with 'reality' television, which includes the talk television of Jerry Springer or Esther Ranzen, as well as the reconstructions of shows like *999*, and the participation of the public in, for example, *The Antiques Roadshow*, *Points of View* and *Changing Rooms*. Reality programmes focus on the lives of ordinary people and depend on them to function as subjects, participants and presenters. (*The Matchmaker*, BBC 1, 1999, uses its main subject both as on-screen and voice over narrator.)

It is useful at this point to look at what programmes are categorised as docusoap, before going on to explore its formal structure and appeal for both audiences and broad-casters. According to a survey in *Broadcast* (Phillips, 9th October, 1998), the highest viewing figures between September 1997 and August 1998 for popular factual series went to *Airline* on ITV, an observational series about the employees of Britannia Airways (chang-ing to Easy-Jet in the second series) which received an average audience of 11.48 million, or a 50 per cent audience share. This was followed by *The Cruise* (BBC1, 1998) with 10.39 million viewers (41 per cent audience share) and *Holiday Reps* (BBC 1, 1997), with 9.60 million viewers, 39 per cent audience share. If there is a conclusion to draw from this survey, it is that our leisure pursuits seem to appeal most to both programme-makers and viewers – holidays (also including programmes such as BBC's *Airport*, 1st series BBC 1, 1996 and *Pleasure Beach,* BBC 1, 1998) and shopping (*Superstore*, BBC 2, 1998; *The Shop*, BBC 1, 1998; *Lakesiders*, BBC 1, 1998). These interests are also reflected in the large numbers of holiday, fashion and lifestyle programmes which, like docusoaps, fill the schedules in the early evening. Their respectable viewing figures, which at their best match sitcom and game show figures and begin to come close to soap opera ratings, point to the wide audience appeal of these programmes.

While most popular factual series are scheduled in the early evening, some recent docusoaps have been transmitted in post-watershed timeslots: *Jailbirds* (BBC 1,1999, Mondays and Tuesdays at 9.30pm) and *The Matchmaker* (BBC 1, 1999, Fridays at 9.35pm). The more adult audience expected at these later times is reflected in the content of these programmes. *Jailbirds* focused on women in prison and their crimes, drug-taking, mental health and sexuality. The subject of *The Matchmaker* was an introduction agency for the middle-aged and middle class and the relationships of its owner, his staff and their clients. Although I am discussing docusoaps as a fairly cohesive group, differences within the genre exist and may be increasing as the format establishes itself and innovates. I will return to possible future developments at the end of the chapter.

Docusoaps, like other 'reality' programming, enable access to the media for a range of people who would not ordinarily be considered of interest, stewardesses, supermarket cashiers or estate agents. This is the populist argument for docusoap, 'a form of people power' (Ruth Pitt, Head of Documentaries, Granada Television, speaking on, *On Air*, 'The Truth Behind TV: Docusoaps'). For some commentators this access decreases the stranglehold of the middle-class, middle-brow broadcasting establishment and allows for the representation of ordinary people in all their diversity. In this argument the 'real' people in these programmes are not victims of an exploitative medium, but part of a democratising trend, which opens up the media to a wider public.

Alternatively, docusoaps can be read as exposing and manipulating their subjects to feed the audiences' desire for voyeurism and gossip. A concentration on personal revelations is shared by docusoaps and the reality programming of talk television. On the surface the attraction of these programmes seems obvious, a chance to peek into other people's lives in the hope, usually fulfilled, of an emotional, dramatic or ridiculous moment. This both confirms our common humanity as well as comforts us that we, the viewer, are kinder, more intelligent, less 'nerdy' than the subjects of these films. The subjects, as well as the programme-makers, sometimes talk about the therapeutic effects of public confession and exposure. Chris Terrill, director of *Jailbirds*, talking about Toni, one of the programmes subjects, said, 'She's still worried that she might go back on heroin. But I said to her, "Toni, you are now going to have the whole nation on your back" ' (Devlin, 1999). In a world where universal moral guidelines have been destabilised and the institutions of church, family, etc. which policed them, no longer have unquestioned authority, television takes on the role of priest, parent and therapist.

Thus docusoap draws from documentary its actuality, real people in real situations, but with a 1990s concentration on the personal and confessional that is also evident in other reality programming, especially talk television. The self-consciousness with which docusoaps present their subjects allows it to be read ironically by media literate audiences, thus undercutting some of the 'true to life' claims made by other observational formats.

## Narrative and Narration

Much of the concern over docusoaps centres on the narrative construction of its material, which owes more to the rules of drama than to a documentary rhetoric. The debate around the use of drama in documentary is a longstanding one. The drama/documentary divide became institutionalised with the observational mode of documentary in the 1960s, which espoused a purist ethos opposed to re-construction. For the documentary associated with John Grierson in the 1930s and 1940s, drama was not only an important element of the realist repetoire, but a major element in his definition of the form. The 'creative treatment of actuality', as Brian Winston explains, was a call to dramatise the basic actuality material (1995: 99). Grierson was always aware of the need to entertain the audience in order to inform and educate them, and he recognised the power of Hollywood techniques to achieve those ends. Robert Flaherty's film, *Nanook of the North* (1922), to which Grierson first applied the term documentary, is the first dramatised documentary, if not docusoap. Life with Nanook and his family was organised in order to be filmed. Nanook played himself with all the charm and 'hamminess' we now associate with the subjects of docusoap. In today's TV world, Nanook would become a supermarket-opening celebrity with a guest spot on the lottery show. His agent would not have let him starve to death a year after filming.

Although in part a variant of the observational documentary, docusoap also has important links with the dramatic intentions of the Grierson realist mode of documentary. These links include the narrative structuring of its material and the 'performances' of its participants. Docusoaps, like other documentaries, narrativise their material. Firstly, by constructing an overarching chronological framework, they shape and control the amorphousness of actuality. Secondly, by emphasising resolutions for problems raised, they replicate the dramatic structure of fictional narrative. In narrativising their material, documentaries have been known to enhance dramatic elements over the ordinary and the mundane of everyday life. Docusoaps, in their appeal to a wide popular audience, are particularly criticised for controlling and manipulating their material, either through the setting up of sequences, or the juxtapositions of those sequences in the editing. Famous instances include the sequence of Noeline Baker of *Sylvania Waters* (BBC 1, 1993) at the hairdressers during the birth of her grandchild, a sequence which was shot at a different time and inserted to make a comment about her vanity. More recently, the sequence in *Driving School* where Maureen wakes her husband at 4.00 am to have him test her on The Highway Code has been the subject of controversy about the ethics of reconstruction in programmes purporting to be fly-on-the-wall. Whereas the observational documentary has always depended on the claim that the film remained faithful to the time and space of the original event for its authenticity and credibility, docusoaps, in their search for drama, appear more contrived and controlled.

Chronological ordering of material overarches the whole series, as well as individual episodes. *The Cruise* centres around a journey, a common narrative structuring device, while the series, *Jailbirds*, set in a woman's prison, follows a number of prisoners, 'doing their time'. Episode one begins in the morning as new prisoners arrive and are processed, and episode six ends after evening lock-up. Although months have passed in the course of the six episodes, the opening and closing shots of the series reference a typical day. Individual episodes in docusoaps also follow a 'day in the life' structure. This underlines the central focus of most docusoaps, the daily working lives of their subjects.

The other common structuring device, used in both documentary and drama is the problem/resolution format. The locations, main participants, and background are set up in episode one, with some problem or problems introduced which viewers expect will be complicated through the course of the series, but eventually resolved. A common problematic revolves around romance. For example, in *Holiday Reps* the series follows the ups and downs of the reps and their local boyfriends. The arguments and misunderstandings common in relationships are further complicated by cultural differences and family objections. The special edition, *Holiday Reps Get Married,* resolves some of these problems with a wedding, the traditional romantic resolution.

In *The Matchmaker*, the problems also revolve around personal relationships and their ramifications. Alun Jenkins, the main character, wants to break up with his girlfriend,

Sue, but as she is also an employee working for his introduction agency, this is awkward. The questions over the future of their relationship is complicated by the multiple perspectives offered by the docusoap. Thus a simple problem/solution scenario is undercut by insights into both partners' perspectives (through the use of the narrative devices of voice over and direct-to-camera narration by the subjects themselves), as well as by comments from other participants and the overview given by the observational sequences. Like soap opera narratives, the programme provides the possibility for identification with different 'character' positions, as well as providing a more distanced reflection on the personalities, their problems, and how they handle them. This provokes some of the same discussion around episodes characteristic of soap opera reception and problematises the notion of single or simple solutions to the problems raised.

This problem/solution narrative structure often uses editing to draw parallels between associated themes and propose resolutions. Episode 3 of *Jailbirds* concentrates on two prisoners introduced in earlier episodes, Melissa, a 17-year-old drug addict facing theft and fraud charges for the first time, and a 71-year-old woman, Ivy, facing similar charges, who has been in and out of prison. The stories of these two women are intercut to produce comparisons between them, their age, their experience of prison and their offences. In addition, both women have family problems and both are hoping to receive probation rather than a prison sentence. The editing of the sequences highlights these comparisons by privileging thematic connections over strict chronology. This leads to the charge that docusoaps oversimplify complex issues by making easy or common sense connections.

As noted in *The Matchmaker*, voice over narration is an important docusoap convention in providing different perspectives on events and subjects. It also provides information and continuity which might otherwise be established through scripted dialogue in drama. This narration helps to fill in gaps, introduce the participants, explain their actions and motives, and trail the content of the next instalment. The scripted narration is instrumental in drawing out themes and meanings for the viewer. It can also mark the difference between the more entertainment-based docusoaps and those with more serious documentary pretensions. Thus *Jailbirds* uses the voice of the film-maker to increase its authority as authored documentary, while *Pleasure Beach* (BBC 1, 1998) uses the voice of Nick Hancock and *The Zookeepers* (BBC 1, 1st series 1998) uses the voice of Richard Wilson. These performers, celebrities in television comedy and light entertainment, mark these series as part of television's entertainment output. Their commentary also tends to be nearly continuous over the images, with their 'patter' drawing out the humour and irony of the events viewed. This has much in common with a comic turn or variety monologue, giving docusoaps their lighter, more entertainment orientation.

Although the narrativisation of documentary through the use of chronology and problem/solution formats to organise actuality material is common, it is the incorporation of soap opera narrative strategies which mark docusoaps as a hybridised format. Docusoaps employ soap opera narrative conventions, with their emphasis on multiple characters and multiple storylines which intersect within individual episodes and carry across episodes, to engage and entertain viewers. Soap operas, as television's foremost contribution to popular drama (Geraghty: 1991), consistently top the television ratings with the result that aspects of the form, particularly its narrative devices, have been incorporated into other television drama formats. Robin Nelson (1997:24) calls these narrative devices 'flexi-narrative'. Flexi-narratives combine the on-going multiple story-lines of soap opera with the fast paced editing style and short segments characteristic of a 1990s television aesthetic.

Docusoap series introduce a number of subjects connected by their work or environ-ment. Thus the opening sequences mirror similar sequences in soap operas, shots of place and characters who are the central focus of the programmes. The stories of these characters/subjects alternate and intersect throughout the episodes. Pace is provided by the speed of cutting between the segments. In Episode 3 of *Jailbirds*, an interview between a young prisoner, Melissa, and a drugs counsellor at the prison is intercut with an interview between her parents and the film-maker at their home. Of the eleven segments which make up the seven minute sequence, only one is longer than a minute. Thus two fairly static sequences of interviews gain pace and drama through editing.

As well as multiple storylines and a pacey delivery, docusoaps also utilise a serial narrative format, which has become ubiquitous in television drama. As a device for attracting and maintaining audience interest, seriality has become a standard strategy in television drama. In soap opera this seriality is continuous, without end, although partic-ular storylines begin and are resolved over a number of episodes, the overarching narrative trajectory only ends for a particular character with death or departure. Serial narratives, using delayed resolutions and cliffhanger endings, can hook viewers. While drawing viewers in, seriality, when combined with scheduling several episodes a week, also functions to create familiarity with characters and situations. This familiarity and continuity, scheduling several episodes in a week works to replicate the real time of life experienced by the viewer, combines with settings which, although fictional have connections to real places, London's East End or Manchester, and which all enhance the realism which is so important to British soap opera.

Similarly docusoaps employ the familiarity and continuity provided by a serial struc-ture to build and maintain their audience. Several docusoaps have begun to mimic the soap opera device of scheduling several episodes per week. Both *Paddington Green* (1999) and *Jailbirds* have been broadcast weekly on BBC1 on both Monday and Tuesday evenings. *Paddington Green* has innovated the format further, by broadcasting two ten minute segments on each of the two evenings, which replay particularly popular story-lines, for example, the acceptance of two local girls into drama school and their auditions and subsequent roles in the West End musical Annie. Docusoaps draw their realist credentials from their *verité* aesthetics, actual people and places filmed in a style which promotes a sense of life captured as it is lived (hand-held shots, natural sound and light, cramped camera positions). But the soap opera emphasis on replicating real time through seriality and scheduling and its emphasis on realistic settings are also important to the conventions of docusoaps.

The choice of possible settings for docu-soaps seems limitless, for example, airports, local authority clamping teams and environmental health inspectors, estate agencies, hospitals and police stations have all been docusoaped. These locations provide a space in which interaction between the subjects and/or between the subjects and the public can occur, as well as providing continuity and coherence between the various stories told within the environment. As with most dramatic narratives, the interaction hoped for (and planned for, in terms of who are chosen as the main participants) is one of conflict.

Although the workplace locations might give the sense of a 'behind the scenes' look at some of the services with which the public and therefore the viewer engages regularly, the programmes rarely give much information about the workings of these institutions. In this sense they replicate a dramatic ethos rather than the information ethos of documentaries. It seems unlikely that a desire to know more about airports or supermarkets or estate agencies would underlie repeated viewing of the programmes. Even the makers of docusoaps agree, 'In the end, it's not about locations, it's about characters' (Waller: 1999). In docusoaps the subjects are central in much the same way as in fictional drama, that is, they motivate action and are affected by and develop as a result of those actions. In documentary, much of the participants' backgrounds and personalities are ignored and only those things which are relevant to the issues under examination will be focused on. Whereas, in docusoaps the subject is the focus, and issues such as work, family and relationships are important only in so far as they explore how these individuals cope with life problems.

## Character and Performance

As Christine Geraghty notes in her discussion of Charlotte Brunsdon's study of the British soap opera *Crossroads*, '…the question determining a soap opera narrative is not "What will happen next?" but "What kind of person is this?"' (1991:46). Similarly, Robin Nelson, in discussing the new flexi-narrative format for drama, states that the shift towards increased concentration on character has been at the expense of action. Allied to this shift is a privileging of performance – '…performance alternatively conceived, that is, in terms of a display of celebrity status or lifestyle' (1997:23). This seems equally true for docusoap, where knowledge of the world is subordinated to knowledge of the 'character'. Some of those 'characters' have become minor celebrities. The interchangeability of the terms character (in fiction) and subject (in documentary) again highlights the hybridisation associated with docusoap. The performance of the docusoap 'star', 'hamminess' and 'acting up' to the camera, is a distinguishing feature of the genre.

Participants fulfil a number of functions in the docusoap. They serve as characters in the drama, playing themselves in their work or home environment. As such, they act and interact in the same way as fictional characters in dramas or as subjects in direct cinema – behaving as if the camera were not there. But, at other times, they perform as presenters and experts, guiding the viewer around their world and commenting authoritatively on it. This is part of the populist notion associated with reality television – ordinary people not only being observed, but making observations.

Jeremy Spake in *Airport* deals with the crises that arise in his job as an Aeroflot supervisor, while continuing his amusing commentary on the situations. Thus he is one of the subjects of the programme as well as a presenter and expert. The BBC series *The Matchmaker*, discussed previously, is more unusual in terms of the narrative roles taken up by its subject, Alun Jenkins, the owner of the introduction agency. At times, he is filmed in an observational manner participating in events. At other times he comments on events either directly to camera or through voice-over. In most direct-to-camera pieces in docusoaps, there is an awareness that the camera and crew are there, either through off-screen questions which are heard by the viewer, a form of interview, or by the behaviour of a subject obviously interacting with the film crew rather than directly addressing the viewer.

But in *The Matchmaker*, some of these direct-to-camera pieces take on the quality of video diary entries, a format in which the subject has authorial control over the content of the film. This authorial position is emphasised through Alun's use of the past tense in some of the sequences, thus appearing to give him knowledge and control over the events that are to follow. This authorial position is further emphasised by the lack of any voice-over by either the director or another media personality. On the one hand, the privileging of Alun's voice gives him a position of authority in the text. But on the other, this mixing of narrational techniques results in a plurality of perspectives from which the viewer can assess Alun's self-knowledge and reliability and, ultimately, judge his version of events.

Alun, with his odious views on women and romance, also highlights the fact that, although the ordinary and everyday are said to lie at the heart of soap and docusoap, the choice of participants favours the eccentric and the unusual. These docusoap subjects often reference common television stereotypes. As David Aaronovitch comments in the documentary 'The Truth Behind TV' (BBC, *On Air,* 1998): 'The trick is to find a work-place and pick the 6 characters who most closely resemble the cast of *Are You Being Served* – you know, a sexy one, a 'shouty' one, a camp one, a straight man or animal and then follow their individual crises and triumphs.' These stereotyped central characters in docusoaps have been a source of much of the abuse levelled at the format.

Stereotypes have long served the needs of drama, by providing characters that are instantly recognisable. Television, with its constant interruptions and distracted viewing, has been theorised as more dependent on stereotyping and repetition to maintain narrative coherence and keep audience attention. Hence the charge that all television is overly formulaic. Certain types, as the Aaronovitch comment highlights, seem to be over-represented in docusoaps. They include middle-aged, working class women, either independent working women such as Eileen from *Hotel* or slightly incompetent and comic figures like Maureen (*Driving School*). These types are reminiscent of the women characters in soap opera. The other familiar character is the camp male, shop assistants, local authority clampers, airline stewards, waiters, who resembles stereotypes from popular situation comedies.

But although stereotypes are important to docusoaps, the familiarity engendered by seriality and the realism of everyday 'crises and triumphs', makes these people more complex and knowable than mere comic stereotypes. Thus, while Alun Jenkins might be a character we love to hate because of his '…patronising, exploitative and manipulative… pontifications about the nature of relationships' ('Watch This', *The Guardian*, 23.4.99), seeing him make a fool of himself, being banished from his own party and expressing his feelings of contrition to the camera (*The Matchmaker*, Episode 2) gives us some understanding of him, if not empathy.

The concentration on women and gay characters in docusoaps also points to the importance of personal and emotional issues in these programmes. Traditionally these issues have been most closely associated with women and their exploration through talk, so central to the docusoap aesthetic, replicates women's pleasure in discussion and gossip. Soap operas have been theorised as the part of television production that is primarily addressed to women through their scheduling in the early evening, their focus on the personal, and the strong female roles at their centre. This is the position taken by Geraghty (1991) in her discussions of soap opera and by Shattuc in her book on daytime talk shows (1997). Both see these formats as a celebration of women's friendships and competencies.

This theoretical approach to the gendering of the audience has become more problematised in contemporary discussions of media. The boundaries between programming that was traditionally considered male and traditionally considered female have become more fluid. The personal and the emotional have become as essential an ingredient in police series as in soap opera, while soaps have increased the number of central male roles and incorporated more action-oriented storylines. Part of the explanation for these changes is the attempt to appeal to as wide an audience as possible when the audience is perceived as increasingly fragmented. In the docusoap, incorporating a variety of characters in the programmes of different ages, genders, sexual preferences and ethnic backgrounds provides a range of potential viewpoints, a range of potential storylines and the possibility of a diverse set of identifications between audience and 'characters'.

Although there are a large number of women who feature in docusoaps, many participants who have received attention both from viewers and the media, are male. These tend to be camp figures (Ray from *The Clampers*) or sexually ambivalent ones (Jackie in *Paddington Green*). The focus on camp characters represents the incorporation of a camp aesthetic in docusoaps, which undercuts some of the realism associated with documentary and British soap opera. Whereas the realist mode references authenticity and seriousness, camp represents a celebration of play, humour, and irony. The emphasis on performance and exhibitionism, common to both male and female participants in docusoaps, is part of the spectacle of camp.

Spectacle is also referenced in elements of the visual style. The wobbly, hand-held shots and the indistinct sound associated with *verité*, which stands for unmediated reality, are juxtaposed with a more 'cinematically' flamboyant style. Again, *The Matchmaker* has taken some of this play with style to an extreme. As well as focusing on a central character who can be described as excessive in terms of his views, his demeanour and his appearance, the series has also presented him through a mixture of styles. Some of its sequences are shot with the awkwardness associated with unscripted actuality. But there are also highly elaborate aerial shots, which exceed any narrative or informational purpose and draw attention to themselves as spectacle. This pastiche of styles in the docusoap, the naturalism of documentary combined with visual spectacle, opens up the text to ironic readings, which are further underlined by the emphasis on performance and the celebrity status achieved by some docusoap 'stars'. Thus the seriousness of their observations of ordinary life are deflated by the playful, ironic style in which they are presented.

## Some Concluding Remarks

Docusoap provokes heated argument that the genre is destroying serious documentary. This view is based on the docusoap's concentration on personal, 'lifestyle' issues delivered through an entertainment package of popular drama conventions, eccentric characters, and a pastiche of visual styles all of which problematise its documentary credibility. Its appeal for audiences, though, may be just this hybridisation. Just as these programmes utilise a range of tactics to appeal to a diverse audience, so the responses to them may be equally diverse. One might respond to and identify with the serious issues of coping with everyday life experienced by the participants. Equally one might read in the programmes' excessiveness and flamboyance an irony that deflates the 'seriousness' of documentary, opening up possibilities for pleasure purely in entertainment and diversion. Such a reading could also be defined as progressive, in as much as irony unseats the moral pretentiousness of much 'quality' documentary and drama. Another possible response to some or all of these programmes is one of cynicism. This response sees docusoaps as examples of programming appealing to the lowest common denominator, voyeuristic, exploitative television whose only purpose is ratings success.

But rather than being exploited by television, it seems increasingly apparent that the 'stars' of docusoaps are using television for their own ends, rather than being used by it. Television seems increasingly to be a site of competition between programme-makers and participants for control of the medium. As documentary film-maker, Phil Agland, whose own recent films, *Beyond the Clouds* (Channel 4, 1994) and *Shanghai Vice* (Channel 4, 1999) have more than a passing relationship to docusoap, has said, '…the whole genre of British docusoap offers an insight into late 20th-century televisual culture…. In 20 years time, we might look back at that and say they are part of a media-conscious culture' (Montgomery, 1999).

## Bibliography

Baudrillard, J., *Simulations*, New York, Semiotext(e), 1983.

Dams, T., 'Time to Move On', *Broadcast*, 23 October 1998.

Devlin, A., 'On Drugs and Behind Bars', *The Guardian*, 11 March 1999.

Feuer, J., 'Genre Study and Television', in Robert C. Allen (ed.) *Channels of Discourse, Reassembled*, London, Routledge, 1992.

Geraghty, C., *Women and Soap Opera: A Study of Primetime Soaps*, Cambridge, Polity Press, 1991.

Hamilton, A., 'Brain Surgeons from Hell', *Journal of the Royal Television Society*, 35:7, October 1998.

Kilborn, R. & Izod, J., *An Introduction to Television Documentary: Confronting Reality*, Manchester, Manchester University Press, 1997.

Montgomery, I., 'Welcome to China, *The Guardian*, 26 February 1999.

Nelson, R., *TV Drama in Transition: Forms, Values and Cultural Change*, London, Macmillan Press, 1997.

*On Air*, 'The Truth Behind TV' (producers: Mentorn Barraclough Carey, series dir: Kate Williams, presenter: David Aaronovitch) BBC2, 1998.

Phillips, W., 'Pop Fact Fabulous', *Broadcast*, 9 October 1998.

Shattuc, J.M., *The Talking Cure: TV Talk Shows and Women*, New York, Routledge, 1997.

'Watch This', *The Guardian*, 23 April 1999.

Winston, B., *Claiming the Real*, London, BFI, 1995.

# What's All This Then?: The Ideology of Identity in *The Cops*

## Madeleine K. MacMurraugh-Kavanagh

## Introduction

Focusing on the recent television drama series, *The Cops* (1998), this chapter explores the ideology of identity in relation to the series' reinscription of the generic conventions of TV police drama and of the iconography of its central characters. The discussion approaches this issue from three central points of view: firstly, the argument indicates the ways in which *The Cops* interrogates the hegemonic logic of police-drama by proposing a new identity for the genre complete with formal innovations which nominate traditional narrative form as an ideological intervention; secondly, the means by which the series converts the identity of the police characters from 'heroes' into a site of anxiety and ambiguity is explored; thirdly, the debate examines the audience's negotiation with meaning in a context where its ideologically-generated belief-system and socio-political identity is consistently challenged. Based on original and previously unpublished material, this discussion therefore proposes that in relation to genre, representation, and audience, the ideology of identity provides the analytical key to this groundbreaking series.

When *The Cops* hit the screens on 19 October 1998 (BBC2), the conventional intersection between hegemonic discourse and police-drama narrative was shattered. Traditional television 'cop shows' achieve meaning through symbiosis between ideological assumption and dramatic structure – the 'beginning' involves a disruption to the status quo by a 'deviant' agency (the villain); the 'middle' is occupied with police procedural as the villain is identified by a usually 'maverick' detective hero; resolution involves the apprehension of the villain, the neutralisation of the threat he poses to the status quo (coded here as 'law and order'), and, since the audience is in tacit agreement that he is justly ejected from 'legitimate' society, the verification of that society as 'right-thinking' and valid. As a result of this conventional structure, which occurs so regularly that it has become a formula, a binary movement between ideological message and 'cop show' narrativity is produced; from it the audience extracts comfort, meaning, socio-cultural orientation, and socio-political identity where a legitimate 'us' is defined in relation to a deviant 'them'. In such a model, a chain of dichotomies articulating a hegemonic value-system based on oppositions including 'right/wrong' and 'order/anarchy' coerces, controls, and contains the audience in the very act of its spectatorship.

*The Cops*, however, refused this pattern of hegemonic meaning-generation, and offered in its place a wholescale reinscription of the genre to which the series ostensibly belonged. With Tony Garnett as its Executive Producer, *The Cops* radically revised the emphasis, grammar, and structure of the TV 'cop show', analysed the identity of its police characters in an appropriation and exploration of 'stock' roles, and, in doing so, rewrote the socio-political agenda of this most popular television genre. This discussion explores the means by which this reinscription is achieved, connects the agenda it announces with a series of ideological and dramatic question-marks, and argues that all the formal innovations and socio-political anxieties central to the series indicate the presence of an ideology of identity that consciously inverts the generic and representational rules of the 'cop show' to produce a drama of sustained interrogation.

Every formal, aesthetic and political revision involved in this series is informed by a concern with the problematic implications of identity as targeted in three dominant strands of analytical activity. Firstly, *The Cops*, its title plugging into an entire televisual history of police-drama, examines the formal and ideological properties of the genre and rewrites both – as a result, the title may invite generic designation, but the series systematically shatters that alignment and simultaneously subjects the whole identity of the 'cop show', together with its structure of assumption, to critical enquiry. Secondly, the series utilises a chain of 'stock characters' in order to analyse not simply these identities, but also to expand and develop the audience's understanding of the socio-political genesis of its response to them – these characters also raise questions regarding the absurdly heroic notion of TV 'cops', substituting in its place a vision where the identity of the police characters is complex, morally-fraught, and ambiguous. Thirdly, *The Cops* targets the identity of its audience, a viewership 'preprogrammed' to endorse the familiar hero-detective and belief-system of police-drama, and subjected here to a process of radical disorientatation in relation to both. In each of these manoeuvres, *The Cops* connects the issue of identity with an agenda of systematic ideological scrutiny.

Given Tony Garnett's track-record in television production, it surprised no-one that *The Cops* proved to be controversial, groundbreaking and powerfully critical. From *Cathy Come Home* (1966), through *Days of Hope* (1975), *Law and Order* (1978), and recently in *Between Lines* (1992 – 4), Garnett's unwavering concern with the socio-political fabric of contemporary society has produced a chain of social realist dramas where the nature of authoritarian institutions, and the collision between authority and the individual, become centralised themes. *Law and Order* proved to be a particularly confrontational contribution to this debate, arousing 'the ire... of the Prison Officer's Association and the Metropolitan Police'; the series was subsequently refused a repeat run on British television and was denied overseas sales (Brandt, 1981,p.30). The drama focused upon police corruption, a theme that Garnett later developed via the 'cynical realism' of *Between the Lines*.[1] But where he describes his earlier work (including the 'Wednesday Play', *Cathy Come Home*) as 'just a howl of rage', indicating clear problems with clear answers, Garnett now admits that he is no longer 'even sure whether I know all the questions' (interview, 5 January 1999). As a result, a growing moral and philosophical complexity characterises Garnett's more recent work where the answers to ambiguous situations are as elusive as the questions. *The Cops* is a product of this ethical density and, in its desire to 'lead people into seeing how complicated the problem is' (interview, 5 January 1999), refuses knee-jerk response and easy analysis at every turn.

One of the ways in which this complexity is registered in Garnett's latest series concerns the issue of generic identity. *The Cops* is clearly not what it seems to be – announcing itself through its title as a police-drama, it immediately becomes obvious that 'the cops' here are merely a point of entry into an altogether more pressing dramatic context. Indicating that he had no particular desire to become involved in making yet another television police-drama, Garnett states that his interest in this series derived from a concern to 'bear witness' to the range of socio-political problems that scar 'the other side of Blair's Paradise'[2] – these problems include the poverty and despair of 'sink estates', the loss of family and community structures underpinning them, and the sense of hopelessness experienced by those consigned to them. This is a disenfranchised society where ambitions of employment are at best a memory and at worst a bad joke – it is also a society that is rarely seen on television, and never centralised dramatically in the unsentimental terms of its true reality. It was this 'invisible' landscape that Garnett wanted to explore in a new television drama, but experience suggested that any format allowing for entry into this context would be met with stern resistance both by broadcasting institutions and also by viewers – 'in these right-wing times that like to think of themselves as apolitical', to produce a series which represented this level of social despair would have generated a drama that was simply too 'on the nose, too direct: you'd never find an audience' (interview, 5 January 1999).

The solution was to smuggle this context onto the television screen via 'trojan horse drama' (Hill, p.156).[3] A trojan horse (something other from what it seems to be) is designed to penetrate a fortressed system – apparently 'harmless', it in fact contains a lethal capacity for destabilisation. By producing a 'cop drama' located in a fictional Northern town big enough to experience urban blight but small enough to allow for the development of relationships between police and policed, Garnett knew that the institutions would commission since an audience would undoubtedly, given the popularity of police-drama, be persuaded to watch. The focus upon uniformed officers rather than the usual detectives would allow the drama to access the petty-crime ridden context of an underprivileged estate thus leading a potentially resistant audience to acknowledge the existence of 'parts of society that are not shown on television and that are not the experience of the middle class viewer' (seminar, 24 November 1998). The ideological trade-off involved in this transaction is clear. The strategy recalls a precedent in the BBC's groundbreaking *Z-Cars* where John McGrath (director) and Troy Kennedy Martin (writer), particularly during the first series (1962), used the police-drama format to access the context of a wider community at the level of daily experience. As in the formative drama, *The Cops* centralises this 'felt' 'reality' in a dramatic impulse that admits a 'level of latent critical social analysis' into its narrative structure, and, in doing so, releases a sequence of available meanings (Laing, 1991, p.134).

Garnett, however, perceived a central difficulty in representing this context to derive from the opposition between 'fact' (roughly aligned with 'truth') and 'fiction' (roughly aligned with 'fabrication') implicit within the characteristic flow of contemporary television. Arguing that the bulk of programmes are now, to use a questionable term, 'factual' (news, current-affairs, documentary, sports broadcasting, even the chat-show format), and noting that 'the techniques of factual programming have moved further and further away from the way that fiction is traditionally shot and edited', Garnett states that a situation has arisen where 'almost every drama *stands out* from the flow; you know from the way that it's structured that you're watching "a drama"' (seminar, 24 November 1998). In other words, 'drama' identifies (and nominates) itself as 'other'. Such a situation involves an obvious dilemma for the social realist film-maker since, if factual television is received by the audience as 'real' and 'true', 'drama' is marked out by this audience as 'unreal' and thus 'untrue'. Committed to exposing the 'other side of Blair's Paradise' to a middle-class audience, Garnett had no desire to see that context categorised as anything other than brutally truthful – as such, the conventions of 'drama' ('fabrication') had necessarily to be radically revised in the interests of communicating this context in *The Cops*, and in the interests of extending the potential of social realist drama in the contemporary context.

In effect, this revision aimed to create a dramatic representation that developed the work of drama-documentary (and even docusoap) in bridging the gap between fact and fiction, and thus between 'truth' and 'fabrication' – with the aim of creating the sense of a 'lived' and 'felt' 'reality', all the formal experiments into form and style in this series stem from this drive. For example, where the conventional police drama 'sets a hare running at the top of the show and catches it at the end', *The Cops* 'buries' the 'hare' within the flow of event in order to persuade the audience of a sense of 'reality' (seminar, 24 November 1998). Garnett acknowledges that a more accurate 'reality' would be reflected if the 'hare' was never caught at all; however, he is also aware that to eradicate narrative line entirely would 'completely lose the audience', so a compromise is here enacted where an impression of the 'feel' of reality is enshrined within the narrative properties of the series (seminar, 24 November 1998).

As such, *The Cops* dispenses with the conventional 'beginning, middle, end' structure of police-drama (together with its hegemonic logic), and aims instead to create the sense that it is following a group of people through an eight-hour shift; long stretches of boredom and inactivity are punctuated with bursts of event, no conventional 'beginning' is offered and closure is resisted so that, as in Garnett's early work (see, for example, the 'Wednesday Play' *Up the Junction*, 1965), what the audience finds is an extended and ongoing 'middle'. Other than a crackle of police radio communication over the closing credits, no superimposed soundtrack disturbs this impression of an observed 'slice of life', and an 'open' text which refuses to dismiss ongoing socio-political questions results.

The most crucial of these formal experiments, however, involves the decision to dispose with traditional two-axes film-making with the result that 'the show is almost wholly shot on one axis as though what was being enacted was an unrepeatable event' (seminar, 24 November 1998). Here, *The Cops* makes a clear entry into the conventions of television documentary and appropriates its sense of immediacy and veracity.[4] Related to this is the use of hand-held cameras throughout the series. Garnett explains that the camera-operators are issued with instructions to keep the image as steady as possible, whilst they are also warned that they must react quickly to unexpected events; the resulting images are therefore 'a little rough around the edges occasionally' (seminar, 24 November 1998), and this again contributes to the sense of an unmediated and reactive documentary 'reality' (as well as to a sense of tension and breathless pace). By such means, the project of social realism is extended to its apparent limit.

In its formal revision of the police-drama genre, then, *The Cops* is hardly a 'cop show' at all – it is a dramatic and ideological intervention generated by a desire to 'bear witness' to an invisible section of society in the terms in which it is 'lived' and 'felt'. But if the experience and 'look' of reality is a key priority of the series, as is the formal/ideological challenge to generic identity implicit in this, then *The Cops* is also centrally concerned to deconstruct an understanding of 'reality' that, for the majority of the audience, resides at an unconscious level. Here, the comforting ideologically-generated iconography of the genre is shattered as the audience is brought face-to-face with both its own anarchic fantasies and with its worst fears (socio-political anarchy caused by the criminal 'deviance' of others). Garnett notes that 'most cop shows are about investigations into things that the audience would like to do', and that conventional closure (the criminal is caught) allows viewers to 'feel at ease' with their transgressive impulses (interview, 5 January 1999). Simultaneously, in an implicit acknowledgement of the perception that its viewers' 'comfortable assumptions about affluence, security, personal prestige and power have suffered attritrion ever since the 1960s', conventional police-drama reassures its audience by siting its villains as societal anomalies who will be contained in the resolution of the drama when the crime is inevitably solved (Fiske and Hartley, 1988,p.181); as such, the formulaic 'cop show' allows its audience to enjoy 'the best of both worlds' since it patrols and contains anxieties circulating around 'internal and external law and order' (interview, 5 January 1998).

*The Cops*, however, dispenses with the 'mythological solution' provided by police-drama's unfeasible success in defusing criminal threat, and chooses instead to 'touch the people where they fear, not to frighten them further but to raise political and social questions' (Garnett quoted in Wayne, 1998, p.26). As a result, here we find a hostile and factionalised context where, when serious crime is committed, it is committed by the police more often than by members of the society they seek to control, where situations are too complex to allow an audience unambiguously to vilify, or even to identify, a 'criminal', and where the exact nature of both the divided society and of the structure of authority struggling to control it are subjected to intense scrutiny. This is a context where violence is countered with violence, where no-one is ever clearly 'right' or 'wrong', where 'us and them' formations breed suspicion, confrontation, and obstruction (such as the police versus the estate-dwellers and vice-versa), and where a chain of potential spark-points characterise the nature of lived reality. As such, where the traditional police-drama raises as few questions as possible about the moral and ideological validity of the status quo that it subliminally promotes, and provides any 'answers' in the elimination of all threats to it, *The Cops* proposes whole series of extended questions into the socio-political basis of authority, questions which insist upon a consideration of the identities of its characters and of the contexts to which they belong.

One of the ways in which this interrogatory impulse becomes most evident is in tensions circulating around the identity of the police characters themselves. The title of the series, *The Cops*, refers significantly to a group and no single character is particularly dominant within this group, though the characters of a drug-taking probationer (Mel) and a frequently unscrupulous 'old-school' constable (Roy) have emerged as exciting most Press comment and audience attention. The group, though, consists of several police officers whom the audience can immediately identify as 'stock characters' – Dean (loose cannon, brutal); Natalie (ambitious, tough young officer); Mike (kind-hearted copper); Roy ('traditional', dubious methods but achieves results), and so on. However, having set up these roles, and having allowed the audience to settle into the prejudices generated by them, the series then systematically embarks upon a quasi-Brechtian process of forcing the audience towards reassessment. In the first episode of the first series, for example, Mel is seen snorting cocaine in a nightclub and arriving late for her shift; by the sixth episode, she has emerged as a character whose primary concern is to help the community and who is determined to make a positive difference to the people struggling to survive within it. Similarly, one of the first actions that the audience sees Roy perform is to plant drugs on an innocent party; in later episodes, he feeds a home-less boy on whom he takes pity and emerges as a character whose morality, behaviour and motivation is too complex to either unequivocably approve or roundly condemn.

All of the regular police characters here are taken upon such a journey and, with each, the audience must re-evaluate its initial response to them and acknowledge that its reflex reaction was inadequate in the 'real' context presented onscreen. Unable to rely upon redundant gender or racial stereotypes (Natalie is tougher than her partner, Mike; Jaz resents being ordered to patrol the Asian community and is resented by this commu-nity in turn), challenged and disorientated, the viewer must analyse the identities of these characters, the identity and nature of the context which has produced them and in which they operate, and, crucially, their own identity in relation to the resulting socio-political cocktail. As such, the audience is brought face-to-face with the realisation that 'characters on television are not just just representations of people but are encodings of ideology' (Fiske, 1987,p.9), and is subjected to 'a type of alienation effect so that it can be persuaded to move beyond its fixation on a hero and move instead towards an exami-nation of its own judgements' (interview, 5 January 1999).

In this minefield of ambiguity, of course, 'heroes' are a logical impossibility – this is an ideological landscape that is fundamentally resistant to them and to the hegemonic angle of vision that they encode. As Garnett comments, *The Cops* structurally militates against such a unified viewpoint since it operates through a tension between two points of view – the series is projected through the eyes of the police since we are led into every situation by the police characters, but in terms of editorial position the series is subject to a multi-faceted perspective that gives equal weight to the police and to the 'policed' viewpoint (interview, 5 January 1999). Such an even-handed, binary perspective cannot produce the type of heroes hitherto associated with this genre, and one of the primary attractions of it as far as the majority audience is concerned. Garnett notes:

the audience consists of viewers who ache to root for a particular character. We forbid this because we argue that the situation is more complicated than this would imply; this means that we're withdrawing from the show, and withholding from the audience, the one element of cop-drama that is always expected and can always be provided. It's so easy to provide it, but it would have the effect of making all the things that we're trying to observe and to make sense of serve the purpose of traditional dramatic effect. On this show, we abjure that easy dramatic effect and disappoint the audience in this regard (interview, 5 January 1998).

This is certainly in line with Garnett's consistent desire to attract the 'biggest audience we possibly can but to get it the difficult way, not the easy way' (seminar, 24 November 1998).[5] However, since television drama is definitively heteroglossic and thus polysemic, there are no adequate means of closing down the margins of interpretation (even if this were the aim), and *The Cops* is a series that, to a large extent, is involved in a constant battle with its hero-hungry, ideologically-situated audience. Struggling to neutralise its socio-psychological need to place faith in the values of hegemonic order and to believe in a fundamentally benign notion of authority, this audience may be unable to resist the lure towards simplistic identification with potentially iconic characters. As Hobson notes, textual inflections may invite viewers towards preferred interpretations, but these inflections 'can be changed or "worked on" by the audience as they make their own interpretation of a programme' (Hobson, 1982,p.106). Garnett indicates, for example, that focus groups convened to gauge reaction to the first series produced several responses suggesting that the crooked, morally-ambiguous Roy was the type of officer that the Police Force could use more of since he 'gets the job done and looks after us' (interview, 5 January 1998). The audience is here negotiating a multiplicity of potential meanings and selecting a preferred interpretation where a textual component (Roy's methods) is used to reinforce a distinction between the vague but felt identities of 'us' and 'them' (the questionable means by which he controls 'them' protects 'us'; if he contains 'them', 'we' are at liberty): it is thus refusing the tension between viewpoint and interpretation inscribed within the structure of the series in order to select a familiar, ideologically-resonant meaning. The fact that Roy is seen blurring the boundary between the 'legitimate' and the 'deviant' societies that underpins the hegemonic logic of an 'us and them' distinction, the fact that this causes fundamental problems in identifying and defining 'criminal' transgression, is of no consequence to this audience as long as a reassuring route into identification with the forces of the status quo can be located.

As Fiske reminds us, however, 'all meanings are not equal, nor equally easily activated', and *The Cops* makes this identification impossible to maintain for all but the most tunnel-visioned viewer (Fiske, 1987,p.93). In order to draw attention to the insufficiency of straightforward identification in this context, the series both foregrounds the issue of identity in relation to its regular police characters, and inscribes it into the heart of the narrative: even if the viewer is capable of resisting the text's 'activated' meanings in relation to Roy, Natalie, Mel, and so on, it can hardly fail to resist the meanings implicit in the drama's insistent return to the question of identity at the level of plot. Repeatedly, the problematic nature of identity and the inadequacy of facile, 'automatic' responses to it, is returned to in numerous storylines – these include a conflict concerning the fraudulent identification of an alleged villain by Constables Metcalf and Thomson; a scene in a Magistrates Court where a serial offender is convicted on the strength of his identity despite his innocence, and where the police officer who has framed him is commended on the strength of his coded uniform and 'exemplary' record; a section where Dean removes his numbered epaulets in order to brutalise a crowd of football supports with no fear of subsequent identification; an ongoing mystery involving Sergeant Giffen's shady history; and the identity of Natalie's lover that remains concealed over several episodes (he is finally revealed to be a CID officer in the same station). Further examples are numerous. Each serves to extend the series' nagging concern to centralise the tensions surrounding the issue of identity, and thus to indicate the ideological implications of the hazy notion of authority upon which the audience places such faith. 'Activated' meanings may be resisted by the socially-situated viewer, but when a 'preferred' meaning is woven through several layers of narrative, less disturbing alternatives become increasingly unsustainable. *The Cops*, devoid of a comforting hero and a comfortable hegemonic belief-system, developed upon a series of problematic encounters with the nature of identity, takes its viewers on a difficult journey, and only the most complacent viewers can emerge with their ideological belief-system unchallenged.

In a context where identities, situations and motivations are riddled with complex ambiguities, the police characters of *The Cops* are clearly not heroised ciphers inviting automatic audience identification. However, a less reassuring route into identification is invited by the series – *The Cops*, in forcing the audience to confront the whole range of unanswerable questions and insoluble situations that the police navigate on a daily basis, asks of it, 'if *you* were faced with these situations, what would you do?'; it further suggests that 'a lot of us wouldn't behave any better, and a lot of us would behave a great deal worse', and portrays its characters struggling with human fallibility in a context where superhuman infallibility is demanded of them (interview, 5 January 1999). It is this invitation towards human identification that explodes the mythologised landscape of the conventional 'cop-hero', and it is this recognition that, for Desmond Christy, produces such a 'worrying' vision: 'we are left feeling that when we look at *The Cops* we see ourselves. From such crooked timber, as the philosopher said, no straight thing can ever be fashioned' (Christy, 1998,p.19).

In the emphasis upon the human identity of its characters, *The Cops* makes a further significant departure from the generic rules of conventional police-drama. As Mike Wayne indicates, police-dramas regularly centralise a flawed and 'contradictory' hero who, whilst acting for the 'good' of society, is nevertheless 'clearly separable from the institution of law and order which he serves' (Wayne, 1998,p.31). This hero is too individualistic to tow the official line and is thus a 'maverick'; he may often use methods which are not officially sanctioned and which meet with disapproval from his superiors, but they are nevertheless clearly justified within the drama and are seen to operate in the interests of 'right' in the long-term. As such, both the institution and the hero, though pulling in differing directions, are morally validated within the drama.

In *The Cops*, though, we see a different situation – here, neither the institution nor the group (still pulling in different directions) are morally validated within the drama just as neither are unequivocably condemned. It is not enough to simply argue that deviant agencies here become representative of a deviant agency; rather, the vision of 'mavericks' operating both within and against an essentially 'rightful' institution is here replaced by a vision of individuals whose capabilities and shortcomings are equal expressions of a necessary but flawed model of authority. It is this contiguous articulation that defines the context centralised in *The Cops*, a context that is accessed through a group of complex identities that fracture all confidence in the certainties coded within the conventional iconography of police-drama characterisation. Garnett comments:

> It is not helpful to say, "Aren't our boys and girls in blue wonderful? I won't hear a word said against them", but nor is it helpful to assert that the Police are corrupt, brutal pigs! Neither response is accurate and neither is helpful. Given the fact that we need the Police, then we need to understand the nature of the job and the kind of people who do it. Having said that, the problem that I have with the Police involves the issue of their arbitrary power and the lack of adequate democratic sanctions that are capable of controlling that exercise of power. A situation where the Police investigate themselves is, for me, just not good enough. (interview, 5 January 1999.)

Because *The Cops* deals with a collection of flawed individuals in uniforms, the issue of the arbitrariness of Police power is centralised as a point of serious anxiety – the police are 'human beings, just like us, and they are under particular pressures. But they also have a power over other human beings that most of us don't have' (Garnett quoted in Graham, 1998,p.27).[6] So when the police characters, dressed in the badges of authority which, through hegemonic semiotics, have become associated with a chain of meaning denoting 'order', 'right', 'law', and so on, are seen to commit grave offences including perjury, actual bodily harm, manslaughter, and drug-abuse, when they are seen using their contingent authority to pursue vendettas against members of the public or are seen victimising suspects and securing 'unsafe' convictions, then questions circulating around police abuse of power, and around its lack of accountability, become central. As such, the identity of the police characters in *The Cops* connects directly with a central strand of the series' interrogative socio-political stance since, as in Garnett's *Between the Lines*, the series asks 'the fundamental political question [...] who shall police the police?' (Garnett quoted in Wayne, 1998,p.26).

Since the one 'rotten apple' frequently identified in conventional police drama is here multiplied to produce a whole barrel of 'deviant' officers, this question becomes unavoidable and urgent. But Garnett resists imposing a concrete answer to it as resolutely as he resists knee-jerk reactions to the entire quandrary. His project here is to articulate a range of contemporary anxieties associated with the exercise of power in a society struggling amidst a welter of high-profile corruption cases to resist the unsettling conclusion that the police are 'part of the problem, not part of the solution' (Wayne, 1998,p.38). In rewriting the form of dramatic 'reality', in revising the conventions of 'cop show' identity, in indicating its investment in a structure of hegemonic relationships, and in forcing the audience into a reappraisal of its socio-political identity through a reappraisal of its responses to a range of generic assumptions, *The Cops* presents a context where moral, socio-political, and ideological certainty collapses. The result converts the distinction between anarchy and authority from a 'thin blue line' into a dangerous continuum.

## What's All This Then?: The Ideology of Identity in The Cops

## Notes

1. Tony Garnett, interview by M.K. MacMurraugh-Kavanagh, 5 January 1999. Subsequent references to this interview will be indicated in the main text.

2. Tony Garnett, seminar, University of Reading, 24 November 1998. Subsequent references to this seminar will be indicated in the main text.

3. Tony Garnett used this phase to describe the formal and ideological agenda of *The Cops* both during the seminar he delivered in the series (University of Reading, 24 Novemeber 1998), and during the author's subsequent interview with him (5 January 1999).

4. It is interesting to note that the trailers for a documentary series following a police unit, *Mersey Beat* (BBC2, January 1999), consciously commented upon the 'documentary' tactics of *The Cops*: implicit criticism of the series' perceived transgression into its 'factual' territory was revealed in the trailers' insistence that the 'characters' of *Mersey Beat* are <u>not</u> 'actors' and in its invitation to meet the 'real cops'. The trailers thus seemed to simultaneously pay homage to the persuasiveness of the 'reality' created in the drama series whilst also accusing the audience of being duped by it.

5. Garnett adds that it would be extremely simple to add at least 1 million viewers to *The Cops*' audience of around 3.5 million (including the Saturday night repeated episode): 'first of all, we'd create much more sympathy for the police; we'd turn one or two cops into heroes and involve them in some heroic acts; we'd include a very clear narrative hook in the first copule of minutes and we'd use very clear narratives in each episode […] But I'm not going to do any of these things and luckily BBC2 is not going to make me' (seminar, 24 November 1998).

6. Graham's chosen title, 'Breaking New Ground or Breaking the rules?' fails to recognise that the former necessarily involves the latter.

## Bibliography

Brandt, G.W. (ed.), *British Television Drama*, Cambridge University Press, 1981.

Christy, D., 'Good Cop, Bad Cop', *The Guardian* (G2), 3/12/1998, p 19.

Fiske, J., *Television Culture* London, Routledge, 1987.

Graham, A., 'Shock Tactics: *The Cops:* Breaking New Ground or Breaking The Rules?, *Radio Times*, 24-30/10/1998, pp 27-8.

Hill, J., 'British Television and Film: The Making of a Relationship', in Hill, J., and McLoone, M. (eds), *Big Picture, Small Screen: The Relations between Film and Television* London, John Libbey,

(Academia Research Monograph 16).

Hobson, D., *Crossroads: The Drama of Soap Opera* London, Methuen, 1982.

Laing, S., 'Banging in Some Reality: The Original *Z-Cars*', in Corner, J. (ed.), *Popular Television in Britain* London, BFI, 1991.

Wayne, M., 'Counter-Hegemonic Strategies in *Between the Lines*, in Wayne, M., (ed.), *Dissident Voices: The Politics of Television and Cultural Change* London, Pluto Press, 1998.

# Representation and Reading:

# The Slipperiness of Gender Identity

## Medicated Soap: The Woman Doctor in Television Medical Drama

Deborah Philips

### Introduction

This chapter traces the representation of the woman medical professional in British and American television serial drama. It argues that while the nurse offers the construction of a professional and skilled femininity, it is one that can be easily reconciled within the traditional expectations of the feminine as nurturing and subservient. The woman doctor, however, presents as a much more challenging version of the professional woman. Early television hospital dramas established the doctor as a male hero and ideal citizen, and relegated women to the roles of the supportive nurse or grateful patient. The Mary Tyler Moore company and its range of liberal television series promoted more positive representationsof professional women, and eventually with *St. Elsewhere* the woman doctor became a more prevalent figure in British and American television. Whilst in the 1980s and 1990s the representation of women health professionals has increased, particularly in popular series such as the American *ER* and the British *Casualty*. Nevertheless the woman doctor remains a problematic figure in mainstream TV Drama, such that even prominent and active health professionals are uneasy heroines, frequently shown as physically or metaphorically handicapped in their careers.

The Doctor is a figure who embodies the ideal 'active citizen', the medical professional is a fictional character who can resolve conflicting demands and reconcile contradictory expectations in the contemporary narratives of health. A doctor can be imagined as simultaneously professionally ambitious and community spirited, highly educated yet down to earth, and can combine a familiarity with the most modern medical technology with the traditional skills of care and nurture. And the imaginary doctor of television, whether the real doctor of the chat show or the fictional doctor of the drama series (from Dr Kildare to Dr Greene of *ER*), is among the most trusted and respected figures on television. What is so striking in the last year of the twentieth century is how very few of those doctors, still, are women.

If 1953, and the moment of the coronation of Elizabeth II (Black, 1972, p.169), can be taken as the moment of a mass popular television audience in Britain, this is also the context of post-war reconstruction, and marks a new concern for women as 'active citizens' in the discourses of government and of popular culture. The establishment of a new National Health Service in the post-war Welfare State coincided with a period of new optimism in the field of medical science and technology (Karpf, 1988). The doctor becomes an emblematic figure for the brave new post-war world, represented repeatedly in fiction, film and television (Moody and Hallam, 1998); the 'Woman Doctor' is a figure who would seem to embody the skilled professionalism and citizenship of an idealised post-war femininity and, indeed, she is an established romantic heroine in romantic fiction written by and for a readership of women [1]. But, despite a dramatic increase in post-war applications from women to medical school (Philips and Haywood, 1998) and greater numbers of women doctors in British surgeries and hospitals throughout the 1950s, she is hardly apparent as a figure on television.

In both Britain and America, popular television and cinema representations of the doctor constructed an image of an invariably male authoritarian figure who should not be challenged, 'a handsome paragon of masculine virtue'(Hallam, 1998, p. 32). In the context of the new National Health Service in Britain, the fictional doctor personified an assurance that, as Karpf (1998, p.182) puts it, 'the state could care' and, in America, that the private health care system was benevolent and kind. Such characterisations of the doctor reassured that the hospital and the practice of medicine was in safe (male) hands, in a dramatic convention that has dominated the medical drama since its cinema beginnings in the 1930s (Hallam, p.30). The doctor hero functions in a central dramatic relationship of two men; the hero, an idealistic junior doctor who is subject to the medical knowledge and maturity of a senior figure, so ensuring that both youthful idealism and the wisdom of experience are ensured in the representation of the doctor. Karpf has described this structure in the medical drama series:

> ...the duet of doctors – the older mentor and the Young Turk who clash but together right the world's wrongs often the neophyte doctor was a heart throb, and the older male doctor explicitly a father figure.... But where were the women? They weren't, in the 1960s and 1970s, the doctors.... (Karpf, 1989, p. 189)

Women characters within this structure tend to be positioned as either patients or nurses; the dramatic format of the hospital series is now such an established convention that it has been difficult to write women as anything else. In writing of American television medical drama, Turow notes that:

> In the '50s and '60s, *Medic*, *Dr. Kildare*, *The Eleventh Hour* and *The Breaking Point* were notable for female guest stars playing patients, not physicians. *Dr. Kildare* was especially famous for its procession of dying beauties who fell into the arms of their solicitous, handsome healer. (Turow, 1989, p.175)

It was the handsome Dr Kildare, most remembered in his incarnation as the television actor Richard Chamberlain, who took this formula, already successfully established in cinema, into the television schedules. *Dr. Kildare* became a television series in 1961 (from 28 September 1961 in USA), and was in the top ten television shows until its ratings began to drop in 1966. Kildare had first appeared as the hero of a 1936 film, based on a story by Max Brand, (the author of such macho westerns as *Destry Rides Again*), and by the time he appeared on televison, had already been the hero of fifteen films, a radio series and seven books. *Dr. Kildare* established a structure for the medical drama that was to be continued in the rival American television series, *Marcus Welby M.D.* (1970), and in Britain's *Dr. Finlay's Casebook* (16 August 1962 BBC), and which continues to circulate in contemporary medical dramas[2].

The doctor hero was the dramatic focus of the 1960s medical drama, and, in both British and American medical series, he was supported by a revolving cast of female nurses and patients whose function was to allow for the display of his medical expertise. However medically skilled a nurse may be, her (and in *Dr. Kildare* there is no thought of male nurses) professional and dramatic status is always one of support and servitude to the more highly trained doctor. Dr Kildare was the undisputed hero and the centre of the series, and women nurses or patients were there to provide an inexhaustible supply of romantic possibilities; if a serious attachment threatened to bring narrative closure, the hospital context made it easy to kill her off, a fate that was to befall his fiancée.

*Dr. Kildare* was very protective of its idealised image of the doctor; NBC would not allow the use of the word 'pregnant' and vetoed any story line suggesting anything other than the most respectable of diseases, although Mary Whitehouse (1967, p. 133) still managed to be shocked. Dr Kildare himself embodied the dominance of the doctor in medical care of the time, as Robert N. Wilson has remarked of the American hospital: 'the doctor was not only the central figure in the hospital but a towering one'(quoted in Turow, p.48). The representation of the doctor on American television entirely endorsed that perception; Kildare's mentor and senior tells his male intern in the first episode: 'You know there's nothing special about a doctor. But the attitude towards him is always special.' That 'specialness' and affirmation of the male doctor was to set the tone for the hospital drama; *Marcus Welby M.D.* was to be celebrated by the American Medical Association with a special achievement award in 1970.

The trusted doctor heroes of *Dr. Kildare* and *Marcus Welby M.D.* emerged in a period at which there was a public faith in medical professionals, and in which the medical profession had confidence their own power and technology. As Turow puts it:

> American medicine was powerful, and its leaders had every reason to be sure that their profession would grow even more powerful … The United States was on its way to building a medical research and clinical establishment that dwarfed anything that had come before it. (Turow, p.27)

The British *Emergency Ward Ten* (19 January 1957 ATV) demonstrated a similar confidence and celebration of the contemporary medical establishment and expertise, in the context of the relatively new National Health Service. *Emergency Ward Ten* ran twice weekly and for ten years, from 1957 to 1967; as its title suggests, the drama rested on a medical team rather than an individual hero. It was no less, however, a confirmation of the authority of the male doctor. Praised by the Minister of Health: 'The programme has made a real contribution to saving life and preventing disease' (*ITV 1963*, p.105), and described as a 'Hospital documentary', the medical emphasis was on a new and beneficial technology that was understood by men. Set in the significantly titled town of 'Oxbridge' (at a time when entrance to Oxford and Cambridge colleges was overwhelmingly male), with the exception of one woman doctor, whose dramatic function is as wife to a senior surgeon, women characters once again featured almost entirely as nurses or patients.

*Emergency Ward Ten* was, nonetheless, very much addressed to a female audience; its characters and storylines were novelised by the romance writer Tessa Diamond, and the television series was backed up by a series of Girl's Annuals. The short stories and illustrations of these annuals emulate the front covers of the Doctor and Nurse romances which were so successfully marketed for an older readership in the same period (Philips, 1989, p.140), but the woman doctor heroine of the 1950s and 1960s romance novel does not translate into television terms. The heroines of these cartoon stories confirm Karpf's description of the women characters in the television series: 'sitting behind desks or work-stations, . . . becoming infatuated with doctors – tasks for which the only qualifications are an O level in Make-Up' (Karpf , p.208).

The heroic doctor of the Emergency Ward Ten team was clearly constructed as an ideal husband, rather than as a figure for the female fan to emulate. The first volume of the series of annuals accompanying the series makes it quite clear that a girl's entry into the glamorous world of *Emergency Ward Ten* should be as a nurse; biographies of Nurse Edith Cavell and of Florence Nightingale offer images of nurses as exemplars of English womanly self-sacrifice, but there is no representation of a woman doctor. The nurse is explicitly offered as a model for a contemporary ideal of femininity:

> A Nurse – what a wonderful picture of the best kind of womanhood the word conjures up in one's mind. Cheerfulness … to reassure the sick. Patience and tenderness … to comfort them. Hopefulness … determination … selflessness … to be able to pull them through as far as it is in the power of a nurse to do so. All the attributes that a girl should have – nurse or not – but if she *is* a nurse … the grand opportunity to develop them to the full. (*Emergency Ward Ten Girls' Annual*) 1962, p. 29)

The fictional nurse offers a means of reconciling the new expectations for professional skill and training for women within the traditional attributes of 'femininity'; it is made clear here that the 'power of a nurse' is strictly limited, and subject to the authority of a doctor. While a nurse is by definition in a service role to patients and to doctors, such qualities as 'patience and tenderness' are not so valued in a doctor. A doctor is professionally required to demonstrate qualities of decisiveness and authority, and the woman doctor is expected to have knowledge and intelligence, qualities which are much less comfortably 'feminine'.

If *Emergency Ward Ten* celebrated the medical team, *Dr Finlay* offered a British (or rather, Scottish) television version of the heroic individualistic doctor; like *Dr. Kildare, Dr.*

*Finlay's Casebook* first appeared on television after it had been successful on radio and in the cinema. It ran on television from 1962 (when *Dr. Kildare* commanded a prime slot in the BBC schedule) and continued for eight series until 1971 to become one of the BBC's most successful drama serials; *Dr. Finlay's Casebook* had such a place in the public's affection that it was revived over twenty years after its first showing. Unlike the affirmation of modern medical technology and the new National Health hospital of *Emergency Ward Ten*, *Dr. Finlay's Casebook* was always nostalgic, set in a pre-National Health Service practice in Scotland in the 1920s . The series was based on characters from A.J. Cronin's 'famous stories' of the 1930s (*Radio Times* 1962) and in a rural practice, rather than a hospital. The Doctor hero is a professional gentleman, and a figure at the centre of a community, the medical equivalent of Sergeant Dixon of Dock Green (a progamme running in the same period). Dr Finlay and Dr Cameron ran an emphatically male practice, and their relationship is familiar from the structure of *Dr. Kildare*, with the crusty Andrew Cruickshank passing on the baton of medical wisdom to the handsome and idealistic young Bill Simpson. Women are perceived as a peculiar species (an early episode was entitled 'What Women Will Do'), and the feminine is present only in the resolutely sensible, but medically untrained, housekeeper Janet, and in Dr Finlay's casebook of patients, which supplied, as it had for Dr Kildare, a perpetual possibility of romance.

Dr Finlay was revived by Scottish television (15 March 1993 ITV); this ideal of the male community doctor was enough to survive for three series into 1995. Although the new version of the Doctor now was a demobbed soldier, and his practice part of the new National Health Service of a post-war Britain; the nostalgia and rural setting remained intact, as did the dramatic structure of the relationship between two male doctor heroes. The Controller of Drama who commissioned the revival commented unapologetically:

> We have updated the series without losing the essential ingredients, particularly the relationship between the central characters, which made the original so popular. (Robert Love quoted in Haining, 1994, p. 8)

In *Dr. Kildare*, *Emergency Ward Ten*, and in *Dr. Finlay's Casebook*, the nurses were invariably women, and so were the majority of patients. Medical dramas throughout the 1950s and 60s featured copious film and television representations of the nurse as heroine but the woman doctor is a rare occurrence, and never at the centre of the drama. [3]

It was not until 1978 that America saw its first hospital drama centred on a woman doctor, the little remembered (and not seen outside America), *Julie Farr M.D.*, which was scripted by a woman – but which only survived for six episodes. Two other medical dramas featured women doctors and were similiarly unsuccessful, and as Turow (p.181) comments: 'The fate of these shows along with the abrupt failure of Julie Farr reinforced the prevailing industry attitude that medical series with women as title characters were poison'. [4] The assumption of medicine as masculine on television was not disturbed until the 1970s, although it is still the case that a medical drama centred on a contemporary woman doctor remains to be made.

Women doctors began to creep into hospital dramas in both Britain and America throughout the early 1970s, but this was largely as guest appearances rather than established characters, yet such shows did contribute to familiarising audiences and television networks with the fact that women in medical dramas could play doctors and nurses. In 1970, *The Interns* was the first (if short-lived) American medical series to have an established woman character as one of the interns themselves, rather than as a support in the form of a nurse or administrator. As Turow (p.150) notes: 'The interns they chose reflected the cast of all relevance dramas. There was one black man, one woman, one young married man, and two adamantly single males.' This is a cast shape that does not alter significantly, *Casualty* and *ER*, the most currently successful British and American medical series, have, with slight variations, more or less the same combination of characters. It was the prime time shows of the Mary Tyler Moore production company which established a formula for this kind of ensemble cast, and for the issue-based 'relevance drama', in which women were very much part of the working community. The MTM style has shaped the contemporary medical serial ; the success of their production values enabled issue-based drama to establish itself on the television schedules in both Britain and America.

*The Mary Tyler Moore Show* (19 September 1970 USA) had itself marked a significant shift in the representation of women on television – it broke with the conventions of domestic situation comedy in such programmes as *I Love Lucy* (15 October 1951 USA) and *Bewitched* (1945 USA), in which the woman was the stay-at-home wife [5]. Mary Richards emerged from *The Dick Van Dyke Show* (3/10/61) into her own show. Mary was no longer just a wife, but a working woman, establishing a prototype for the working woman character who could become, in subsequent MTM series, a radio or television reporter, a newspaper journalist, a lawyer or policewoman and, eventually, with *St. Elsewhere* (shown from 26 October 1982 on Channel 4), a doctor. As Bathrick (1984, p.101) points out:

> Mary Richards and her working women friends appear in 1970 as television's first serious concession to a changed world where middle-class daughters leave home, earn their living, and remain single.

The *Mary Tyler Moore Show's* own spin off, *Lou Grant* (1970), established a precedent for a form of serial drama that centred on the workplace, in which the drama and romance emerged from situations at work and women were more firmly established as working players and as central to the ensemble cast that was such a mark of television serial drama throughout the 1980s. *Lou Grant* added an edge of explicit social commentary to the situation comedy of the *Mary Tyler Moore Show*, which had ended in 1977; it represented a new kind of television drama; as Feuer (1984 a, p. 20) explains:

> . . . *Lou Grant* was a transitional programme, the bridge between the Brooks-Burns era of sitcoms and the less comedic dramas to follow. The programme continued the ensemble approach popularised in the sitcoms, while diverging to introduce social issues.

If Dr Finlay and Dr Kildare had established the iconic male doctor heroes of general and hospital practice, in which women remained peripheral to the action, these programmes had also set up conventions for the television drama set in a work context, with a fixed set of characters, ongoing plotlines, and new storylines introduced by patients each week. The pleasures of these series were close to that of the soap opera (Geraghty, 1991), with the resolution of some of the narrative threads, while several storylines remained continuous. The structure of an ensemble cast, featuring a range of character types and set in the work place also had a precedent in situation comedy set outside the home – in Britain, Croft and Perry had familiarised audiences with situations beyond the domestic (Bowes, 1990, p.133). However, the cast and communities represented in these comedies remained resolutely unchallenged by the outside world, and were often set in a nostalgic past. The nature of the work in the MTM drama meant that the nature of the storylines was necessarily contemporary and could be entirely unpredictable, and brought the outside world into the known community of the workplace. Familiar characters were regularly confronted with unfamiliar situations, in a narrative organisation that still structures *Casualty* and *ER* .

MTM went on to produce the police series *Hill Street Blues* (15 January 1981 NBC in USA, shown in Britain from 22 January 1981 by ITV), which, with its hand held cameras and gloomy lighting established a stylistic precedent for socially conscious work-centred dramas, and which was to have, formally and narratively, a great impact on such hospital dramas as the British *Casualty* and the American *ER*. *Hill Street Blues* also identified a new audience demographic and made a network serial drama which took on controversial issues successfully (Feuer, 1984a, p27). Both *Lou Grant* and *Hill Street Blues* marked a new style of drama programming, and established a format in which working women were at the centre of the action, and in which the dramas of their lives were focused on what happened to them at work rather than at home.

The dramatic focus of *Lou Grant*, and of the MTM workplace dramas that followed it, was on a professional team, rather than, as in *Dr. Kildare* and *Dr. Finlay's Casebook*, an individual hero. If *Emergency Ward Ten* had already employed a fixed cast set in a working environment, *Lou Grant* extended that to a wider range of ages, ethnicities and character types. Wicking's (1984, p. 170) description of the *Lou Grant* ensemble remains true for subsequent MTM productions, and is now an identifiable model for the contemporary hospital drama (Feuer, 1984 b, pp. 44-45):

> . . . the archetypal surrogate TV 'family', a cross-section of types and ages corresponding to the main audience groups and containing underlying father/mother/children/lover relationships on a symbolic level . . . the Lou Grant 'family' is closer to the Howard-Hawks-type team of professionals, about whom we're curious . . . wishing to 'understand' them as they go about their work . . .

*Hill Street Blues* developed this 'team of professionals' drama, but it was clearly a format that was ideal for a hospital context, in which the team are inevitably confronted with life and death dramas. In a hospital or surgery setting, every case that comes in has an unpredictable story attached (as viewers of *Casualty* can testify, much of the narrative pleasure derives from attempting to predict which of the characters of the opening story lines will end up in a hospital cubicle). MTM initially experimented with two pilot hospital dramas, *Operating Room* (written by Steve Bochco who was responsible for *Hill Street Blues*) in 1978, was labelled a 'comedy-drama' (as was the later *St. Elsewhere*) but didn't survive beyond the first episode. *Mother and Me, M.D.* followed in 1979, centred around a woman intern doctor who worked at the same hospital as her mother, a nurse. The subject and dramatic tension of the premise does suggest a shift in expectations for women in the 1970s, but it is significant that the pilot never made it into a series.

It was *St. Elsewhere* (26 October 1982 Channel 4) that became MTM's established hospital series. First broadcast as a pilot film in 1982, it survived as a series despite initially low ratings only because of the success of *Hill Street Blues* (which had similarly had initially small audiences), and was directly modelled on Hill Street:

> *St. Elsewhere*, when it was being developed, was referred to around the shop as 'Hill Street in the hospital'. Its style is wholly derivative: the large ensemble cast, the blending of melo-drama and comedy with the more or less 'realist' treatment of the medical series tradition and of controversial issues (AIDS, sex change operations), and in its use of the continuing serial narrative (Feuer, 1984 p.44)

Not unlike the later, and more successful, *ER,* the ensemble cast included a range of sexualities, statuses, and ethnicities, but as Karpf (p.183) has noted of *Emergency Ward Ten*: 'There was a marked absence of porters, ancillaries, black nurses and women doctors.' If the ensemble casts and complex storylines of *St. Elsewhere* did shift the focus of the drama onto characters beyond the doctors, that absence continued (and, with the exception of the improbably patient and sympathetic reception staff of *ER* and *Casualty*, continues to be the case). Among the doctors, there was one token woman, and one token black character (both implausibly beautiful and brilliant). *St. Elsewhere* did feature one nurse character who went on to make the transition from nurse to doctor (a device that was redeployed in the character of George, in *Casualty*, and suggested in *ER*'s Carol) As the series developed, *St. Elsewhere*'s regular cast included among its twelve doctors two women (junior doctors); despite MTM's relative liberalism, it was still the men who were at the centre of the medical dramas, and in positions of authority.

*Casualty* was the BBC's flagship medical drama; first broadcast on 6 September 1986, it has gone to ten series to date, and shows few signs of flagging, in 1996 it had viewing figures of fifteen million (Hallam, p.25) and is still broadcast in a prime time Saturday night slot. In the context of post-war Britain, the fictional hospital stands as an emblem of the Welfare State, and so is also always about the state of England. The fictional Holby General represents any hospital in any town, its mixed race cast representative of a multi-cultural Britain. Since the National Health Service, the medical profession is held in Britain with a measure of real public affection, and the status of the doctor and of the nurse as honourable citizens allows for more politicisation than can be found in most mainstream television drama. Broadcast in the same year as the documentary series *Hospital Watch*, which promised to 'tell it how it is' , February 1986. p. 82), the series was written to celebrate 'the comedy and heroics' (*Radio Times*, September 1986, p.4) of life in the National Health Service. *Casualty* focuses on the commitment and skills of profes-sionals in the face of extreme difficulty, and in the context of Thatcher's Britain, those difficulties were implicitly (and sometimes explicitly) those of government cuts to the health service. *Casualty* regularly points to lack of staffing and resources, and much loved regular and narratively respected characters were heard to question the restructuring of the Health Service in no uncertain terms, to the extent that, in a 1986 Conservative government, *Casualty* prompted questions in the House of Commons (Karpf p.192).

In its life span *Casualty* has had a number of women doctors, of whom the longest standing is the character of 'Baz' (Barbara Samuels), who remains among the most sympathetic women doctor characters on television. Significantly, like her successor, George, her nickname suggested that she, although female, was just one of the boys. Baz was introduced to the series as a student doctor, and developed to become a senior practitioner; her romance with senior nurse Charlie threatened to tip the balance of power and gender relations, and was a constant source of narrative tension. Charlie's masculine prowess was firmly established in an episode in which he rushed to rescue her in a melodramatic encounter with a kidnapper; Baz's professional competence (though constantly narratively stressed) was not enough to make her equal to a male hero. The nurse remains in *Casualty* a more comfortable version of femininity than the woman doctor, as in the character of senior nurse, Kate Wilson who: 'sometimes cares too much and can't help looking after people . . . She is traditional at home and traditional at work' (Kingsley, 1995, p. 65).

*Peak Practice* has also shown the impact of government cutbacks and the restructuring of the health service, and critically represents the effects on a financially pushed rural clinic. First broadcast on 10 May 1993, and produced by one of the *EastEnder's* producers, it has gone to six series and remains a significant part of ITV's scheduling. *Peak Practice* was devised by a woman, Lucy Gannon (who had herself worked as a nurse); Gannon was also responsible for *Bramwell*, broadcast by Carlton in 1995; set in London in 1895 and centred on a woman doctor, it remains to date the only British medical drama series with a woman doctor as the title character. *Peak Practice* has had its share of strong women doctor characters; Beth Glover was the practice's first woman doctor, and was followed by the young and ambitious trainee, Dr Rhiann Lewis, establishing a central female medical presence for the series. The drama had originally centred on the figure of Jack Kerruish, an idealistic doctor in a small rural community , who was very much in the tradition of the A. J. Cronin and Dr Finlay hero. Gannon was nonetheless concerned that the series should not subscribe to an idealised construction of the medical doctor:

> I made sure from the start of *Peak Practice* that I've shown people questioning their doctors. The more pedestals that are knocked away from under these professional people, the better. We, the patients, need to be aware. (Lucy Gannon quoted in Tibballs, 1995, p. 22)

Dr Beth Glover (played by Amanda Burton, who was later to play the terrifyingly professional and academically qualified pathologist in *Silent Witness* (21 February 1996 BBC1) was one of the strongest representations yet seen on British television of a woman doctor, she was an equal (and vocal) partner in the practice, and disrupted the structure of the male relationship established in *Dr. Finlay's Casebook* and *Dr. Kildare.* As Gannon explains:

# Medicated Soap: The Woman Doctor in Television Medical Drama

… I was particularly keen to include a strong female character. . . . I think the good public reaction to Beth Glover vindicated my decision. Having place a female character alongside two men, it was essential to build sexual tension … (Lucy Gannon quoted in Tibballs, p. 9)

Though represented as a strong and competent professional, Dr Glover's emotional life was perpetually troubled throughout her tenure in the series. In a description of anxious feminine professionalism which also applies to Burton's character in *Silent Witness*, the book brought out to accompany the series characterises her thus :

Dr Beth Glover took over The Beeches surgery in Cardale from her father and is the senior partner in the practice. She is an extremely good doctor but a hopeless businesswoman. Her fiery nature brings her into frequent conflict with Jack, with neither being prepared to back down. She is independent and enjoys being in charge of her own destiny but does have a niggling suspicion that she would be even happier with a man in her life (Tibballs, p10)

The American *ER* (like its rival, *Chicago Hope*) was rather glossier than the British equivalents; the doctors and nurses better looking, the equipment shinier and more state-of-the-art, the drama faster paced (and *Casualty* has had to respond to its success). The first series of *ER*, written by former doctor, thriller and screenplay writer Michael Crichton, was launched on 19 September 1994 in the USA, and in Britain on 8 February 1995 by Channel 4. The series opened with a storyline centring on doctors Mark Greene and Doug Ross, who continued to provide the source of the central storylines until the departure of the much-loved George Clooney, who played Ross. The first cast line-up included just one regular woman doctor, second-year resident, Dr Susan Lewis; although there are a number of women surgeons who appear on an infrequent basis in order to demonstrate that there are senior women in the hospital. Although the pilot episode meant her suicide bid to be successful, it is nurse Carol Hathaway who is now the strongest feminine presence and at the moral centre of the drama. The series has been at pains to point out that Carol has the intellectual ability to undertake training as a doctor and that it is only finances that prevent her, but this storyline has not been pursued, and the ambition has not been referred to since.

The women doctors, with the exception of medical students, currently to be seen on *ER* are all literally or metaphorically incapacitated, in what seems to be a displacement of the actual difficulties confronting the woman doctor. Of the qualified women doctors currently practicing in *ER*, one is British (and therefore has to retrain), one, without explanation, walks with a crutch, and another (represented as less professionally ambitious than the others) is HIV positive (infected by her once-promiscuous husband). The most powerful (but physically handicapped) woman doctor on the *ER* is Carrie Weaver, but she is less empathetic than almost any other regular cast member and her personal life is shown to be utterly bleak (she has been allowed one on-screen affair, with a dubious medical salesman). Despite *ER*'s evident modernity in its style and storylines, the woman doctor remains, as an image of independent and educated femininity, an ambivalent television protagonist. As Karpf (p. 208) comments:

. . . women doctors may be problematic to medical drama, but nurses aren't. Drawing on cultural ideas of nursing as a kind of professionalised femaleness, nurses are depicted at best as perpetual geysers of nurturance and intuitive mothers to the world.

By 1975, an idealised version of the nurse was regularly questioned in the serial drama; *Angels* which ran for nine series from 1975 into 1983, was an explicly ironic title and challenged the construct of nursing as traditionally appropriate women's work. Nonetheless, the nurse, whether sanctified in 1950s medical romances (both in film and fiction), sexualised in the Doctor and the Carry On series of the 1950s and 60s, ironised in *Angels* (1975), politicised in *Casualty* (1986), or romanticised in *ER* (1994), remains a much more comfortable heroine than the woman doctor. She is still a romantic focus, and often the moral weight of the contemporary medical drama, as Carol Hathaway is in *ER*. In *Casualty*, Megan Roach is described as 'the Mother Earth of *Casualty*' (Kingsley, p.76).

The woman doctor remains a much more subversive figure than the nurse. Bramwell remains the only British or American central and titled doctor heroine, and she is greatly handicapped by being a Victorian heroine; the constraints on her career can be displaced onto nineteenth-century prejudice. There is currently no woman doctor in a British or American medical television drama who is not somehow represented as handicapped. As Karpf (p.208) has said: 'The Good Doctor, repository of all the hopes we invest in medicine and the ideals we hold dear, remains an invariable component of even the most abrasive medical fiction.' But the good doctor, as a constant cast member and as the focus of the medical drama is, still, rarely a woman.

Television represents the medical profession as the monitors of physical health, and of the national health; is putting an impossible burden on health professionals that they should be not only the guardians of our physical health, but of our moral well being too. But, if that is the function of representations of medical professionals on television, then constructions of femininity in the medical drama have to move beyond the 'best kind of womanhood' of *Emergency Ward Ten*. In the television drama, it continues to be men who are seen at the cutting edge of medical technology, while care in the community remains woman's work.

## Notes

1. For an account of doctor heroines in Mills and Boon romances of the 1950's, see 'White-Coated Girls' in Philips, Deborah and Haywood, Ian. 1998, *Brave New Causes: Post-war popular fictions for women*, London, Cassell.

2. This structure extends beyond the medical drama; it is also to be found in the hugely sucessful *All Creatures Great and Small*,set in a vetinary practice, which ran on British television from 1978 to 1990.

3. The one exception is an American 1956 half hour television pilot, *Kelly,* which starred Larraine Day. Day had featured as Kildare's love interest in the MGM film series of *Dr. Kildaire*, but after her character was killed by a lorry, Day escaped *Dr. Kildaire* to feature as the woman doctor heroine of the title. The fact that the doctor is a woman is the central issue of the pilot episode, but *Kelly* was never developed into a series.

4. A woman doctor as the titled heroine of a medical drama was not attempted again until the 1986 American *Kay O Brien*, which again did not make it past a year.

5. It could be argued that Samantha, the wife of a mere male mortal in *Bewitched*, as a witch with all the power of the underworld available to her, already represented a subversion of the role of women in the genre.

## Bibliography

Bathrick, Serafina, 1984. 'The Mary Tyler Moore Show: Women at Home and at Work' in Feuer, Jane, Kerr, Paul and Vahimagi, Tise, eds. *MTM 'Quality Television'*, pp. 99-131.

Black, Peter, 1972. *TheBiggest Aspidistra in the World.* London: BBC.

Bowes, Mick, 1990. 'Only When I Laugh' in Goodwin, Andrew and Whannel, Garry, eds. *Understanding Television.* London: Routledge, pp. 128-141.

*Emergency Ward 10 – Girls' Annual*, 1962. London: Purnell.

*ER Companion: An unauthorised guide* 1996. London: Signet books .

Feuer, Jane, 1984 a. 'MTM Enterprises: An Overview' in Feuer, Jane, Kerr, Paul and Vahimagi, Tise, eds. *MTM 'Quality Television'*, pp 1-32.

Feuer, Jane, 1984 b in 'The MTM Style' in Feuer, Jane, Kerr, Paul and Vahimagi, Tise, eds, *MTM 'Quality Television*, pp33-60.

Feuer, Jane, Kerr, Paul and Vahimagi, Tise, eds. 1984. *MTM 'Quality Television'* BFI books, London.

Geraghty, Christine, 1991. *Women and Soap Opera: A Study of Prime Time Soaps.* Cambridge: Polity Press.

Hallam, Julia, 1998. 'Gender and Professionalism' in TV's Medical Melodramas' in . *Medical Fictions*, pp 25-47.

Haining, Peter, 1994. *On Call with Dr. Finlay.* London: Boxtree Books.

*ITV 1963 A Comprehensive Guide to Independent Television*, Independent Television Authority, March 1998.

Karpf, Anne, 1988. *Doctoring the Media: The Reporting of Health and Medicine,.* London: Routledge.

Kingsley, Hilary, 1995. *Casualty: The Inside Story.* Harmondsworth: Penguin, Moody, Nickianne and Hallam, Julia, eds. 1998. *Medical Fictions,.* Liverpool: John Moores University.

Philips, Deborah and Haywood, Ian, 1998. *Brave New Causes: Post-war Popular Fictions for Women,* London: Cassell.

Philips, Deborah. 1990 'The Marketing of Moonshine' in Tomlinson, Alan, eds. *Consumption/Identity/Power: Marketing, Meanings and the Packaging of Pleasure,* London: Routledge pp 139-139.

*Radio Times*, August 11-17, 1962.

*Radio Times*, February 15-21, 1986.

*Radio Times* , September 6-12, 1986.

Tibballs, Geoff, 1995. *The Making of Peak Practice.* London: Boxtree.

Turow, Joseph, 1989 *Playing Doctor: Television, Storytelling and Medical Power*. Oxford: Oxford University Press.

Wicking, Christopher, 1984. 'Lou Grant' in Feuer, Jane, Kerr, Paul and Vahimagi, Tise, eds. *MTM 'Quality Televsion'*, pp 166-182.

Whitehouse, Mary, 1967. *Cleaning up TV: From Protest to Participation* , London: Blanford Press.

# Performing (Wo)Manoeuvres: The Progress of Gendering in TV Drama

Robin Nelson

## Introduction

This chapter analyses the increased visibility of women on TV across a range of dramas, with particular attention to those which, more recently, have represented women within professional roles. It considers the gendering of roles not only in connection with narrative drive, but also in terms of a potentially progressive shift from woman as the object of the male look, to woman as the surveyor of the male body. Detailed reference to programmes is framed by an assessment of the conservative effect of the medium itself.

Representations of gender, and especially of women, in popular TV drama series might seem to have come a long way from the days of *Emergency Ward Ten* (ITV, 1957-67) and *Charlie's Angels.* (ITV, 1977-79) Building on the 1980s inclusion of women in active leading roles in series like *Cagney and Lacey* (Orion TV for CBS, 1982-86), *Juliet Bravo* (BBC1, 1980-85), and *Widows* (Euston Films for ITV, 1983), the foregrounding of women's experience is almost commonplace in 1990s TV drama series. Groups of women are frequently central to the given dramatic situation in *Making Out* (BBC1, 1990), *Band of Gold* (Granada, 1995) and *Playing the Field* (BBC1, 1998), and individual women quite regularly take on the role of the male, once defined by Mulvey, as 'the active one in forwarding the story' (1975:12), as in *Prime Suspect* (Granada, 1991), *The Governor* (YTV, 1995), and *Ambassador* (BBC1, 1998). Whether this increased visibility of women amounts to a progressive shift, and in what terms such a progression might now be constructed, however, remains in debate. As the 'reflection theory' of television has long since been called in question, it would be naive to assume that new representations of 'woman'[1] on television simply reflect, or are reflected by, changes to women in real life.

Not only have gender representations shifted, but in theoretical terms, conceptions of the politics of gender have developed from Laura Mulvey's seminal identification of the cinematic role of women 'to be looked at and displayed' (1975:11). Briefly to summarize key elements in the debate, structuralist/semiotic and psychoanalytic criticism of the late 1970s and early 1980s proposed that meanings structured in the codes of the text drew viewers inexorably into specific subject positions. Irrespective of their gender, so the argument ran, spectators were positioned by the structural features of classic narrative in terms of 'the male gaze' with the female as the object of desire. The question of whether or not a female subject position was possible engaged subsequent theorists, particularly with reference to the open-ended narrative form of television's continuous serial, or soap. Disruptions to classic narrative form were seen as having the greatest progressive potential, though the claiming by women of representational spaces typically occupied by men might have held some empowering potential.[2] I shall refer below to these historical senses of progression as set in 'established terms'.

In the past decade the frame of the debate has been re-drawn somewhat. Increasing recognition has been given to the active engagement of viewers in negotiating the meanings and pleasures of the television programmes they watch. Ethnographic audience studies (Morley, 1992) have demonstrated the variant positions taken by different viewers, relative to factors such as their age, gender, ethnicity and class. The general shift from structuralist notions of fixity to post-structuralist conceptions of the slipperiness of language, furthermore, has reconceived the viewing process as one of dialogic engagement between viewer and text rather than the imposition of fixed meanings or immutable viewing positions on the reader by the text. It is in this latter context, with particular reference to the functioning of the language of representation, that I propose to reconsider in this chapter the idea of gender progression in 'new terms'.

My initial interest has been to invoke a sketch progress review of gender representation in popular TV drama series with particular, though not exclusive, reference to women. My main focus hereafter will be an attempt to frame an aspect of the late 1990s in terms of selected examples of women in the medical-crime fictions who are positioned at sites where a range of discourses of gender intersect. My interest, that is to say, is not ultimately in the general 'caring 1990s' representations of men in aprons doing their share of the dishes or the assault on assumptions about women's sexuality of *Sex and the City* (C4, Spring 1999). Nor is my focus the representation of queer and lesbian relationships which has occupied many commentators on contemporary culture. My topic is the shift in representation of women in professional roles in such a way as potentially to call in question enculturated mind-sets of 'maleness' and femaleness' through new conceptions of the functioning of the language of representation.

The specific cases for consideration are Sam, Dana and 'the woman in the Volvo ad'. Dr Sam Ryan (Amanda Burton) is the leading figure in *Silent Witness* (BBC 1, first series 1996), Dana Scully (Gillian Anderson) is the co-equal lead in *The X-Files* (BBC2, first series 1994), whilst the uncredited 'woman in the Volvo ad' is a doctor operating solo in risky territory. Each woman is represented as a free agent, a professional who operates with high proficiency, if not quite ease, in roles characteristically assigned, in both Western cultures and TV representations, to men. Moreover, to achieve this freedom, the subject of each of my examples is inscribed in that Utopian discourse of scientific rationalism which has been characterized often as the masculine preserve of the Enlightenment tradition. They are all medical doctors, and my key TV drama examples are forensic pathologists who evoke a particular mix of the detective and medical genres amenable to today's television genre hybrids. Whilst it is not a key aspect of my analysis here, the commercial context of the television industry which craves 'the same but different' has increasingly combined genres in an effort to draw bigger and more diverse audiences (see Nelson, 1997: 30-49 and 73-88). Hybridization impacts, as I shall show at the end of the chapter, on the gender issues under discussion.

The advertisement for Volvo cars falls outside standard generic categories of TV drama, though the micro-narrative form of many TV ads has long been recognized. Here 'the Volvo ad' offers a microcosmic narrative and representation to bring the key terms of my debate quickly into focus. The advertisement shows a black woman driving a Volvo estate car through mountainous terrain on ice and snow-packed roads to reach a victim of a helicopter crash. Arriving readily on the scene, she effects medical assistance to the injured. Intercut with this action-narrative are close-ups of scientific instruments and X-rays with the woman's face looking alert and efficient. This example is a paradigm of the contemporary woman agent. A number of progressive aspects in 'established terms' is evident and maybe there is a development in 'new terms'.

A woman, and a black woman at that, has been assigned the role of the call-out doctor in an emergency situation in perilous conditions. She is shown as 'master' of the technology, in controlling the Volvo and the situation. She drives the narrative as well as the car. She appears to have hauled the helicopter to safety and is active in releasing the accident victims. Throughout, as underscored by her cool voice over, however, she is rationally detached. Even when reacting quickly under pressure in a 'testing environment', she knows, as the voice over puts it, 'when to use patience and when to use speed'. The images of scientific equipment, brain scans, X-rays, some overlaid over her face, suggest a person with clear sight and penetrating insight, a capable life-saver by rational and technological means.

Four features typifying my examples for discussion may be brought out from this micro-narrative: 1) the representations of women in high-profile professional roles; 2) the power to drive the narrative; 3) the women's specific association with scientific rational, technological aspects of the medical profession; 4) the exchange of the role of women from Mulvey's 'to be looked at-ness', at least in part, to one of surveyor of other (often male) bodies. Having sketched this model, I propose, before discussion of the TV drama examples, to set out those 'new terms' of contemporary theoretical frameworks which may, or may not, be borne out in televisual practices and the processes of reading.

Gendering traditionally has been located in dichotomous thought-patterns. The key male/female binary opposition, that is to say, has been echoed in a series of other binaries: reason/emotion; masculine deduction/feminine intuition; mind/body; public sphere/private sphere and rational detachment/emotional involvement. The male pretence of rational detachment in the pursuit of an objective truth, and its associated power of cool inductive and deductive logic has historically been situated on one side of a binary divide, with female emotionalism embodied (and I use the word advisedly) on the other. When, on the relatively rare occasions, women detectives have featured in television history, for example, they have been constructed as 'other' than the male paradigm. In their leggy glamour, *Charlie's Angels*, exemplified 'to-be-looked-at-ness' and, although dynamic, they deferred to the unseen master-narrator, 'Charlie'. In marked contrast of representation, the eponymous *Miss Marple* solved the mysteries of her English village world through an intuition constructed as idiosyncratic, the slightly dotty spinster aunt in the wise-woman/witch tradition. Though in one sense Miss Marple drove the narratives, she operated informally, always in parallel with the official investigation.

*Cagney and Lacey* serves as a pivotal example in the history of television representations in the detective genre. The series was progressive in 'established terms' in that, at a specific historical moment when the 1970s feminist movement was contesting dominant representations of women, it was the first dramatic programme in television history to star two women, Tyne Daly (Mary Beth Lacey) and, ultimately, Sharon Gless (Christine Cagney). As D'Acci (1987) relates:

> [t]he characters were represented as active subjects of the narrative who solved their own cases both mentally and physically. They were rarely represented as 'woman-in-distress' and virtually never rescued by their male colleagues. As well as functioning in both the public and the private spheres, they were portrayed as active subjects, rarely as objects, of sexual desire.... The actresses and characters were in their mid-thirties and there was a distinct minimisation of glamour at the levels of clothing, hairstyles and make-up.... They were represented as close friends who took a lot of pleasure in one another's company (1987: 204-205).

To select some key points from D'Acci's documentation of relevance to my subseqent argument relating representation to industrial context, the production history of *Cagney and Lacey* is chequered. The idea, originally in film form, was turned down by studios because 'these women aren't soft enough' (cited in D'Acci: 207). MGM wanted to cast Raquel Welch and Ann-Margret. Offering the project six years later as a pilot for a television series, producer Rosenzweig was told by CBS that they would take it as a made-for-TV movie if he cast 'two sexy young actresses'. The movie was made with Loretta Swit (of M*A*S*H fame) as Cagney, and accompanied by 'exploitation advertising' (see D'Acci: 209-10). In its second phase as TV series, the project met similar difficulties. Despite its relative, if fluctuating, ratings success with Meg Foster in the Cagney role, the series was almost cancelled by the network. In a conflict centring on the representation of 'woman', Meg Foster was replaced by Sharon Gless as a Cagney reconstructed with a middle-class background accompanied by a fashion image make-over. As D'Acci summarizes, '[t]he evidence points to an extreme discomfort on the part of the network with 'woman' presented as non-glamorous, feminist, sexually active and working-class and single' (1987: 214).

The issues and tensions in the sketch above show how institutional forces in television have militated against TV representations which blur gender distinctions, preferring clearly distinguished gender 'norms'. It evidences, in other words, a binary mind-set amongst the predominantly male executives in the industry. The industrial stereotype sense of 'woman' as sex object to-be-looked-at inevitably produced constructions for the heterosexual 'male gaze'. When a representation of 'woman' did not conform with these norms, it could only be conceived as 'other', 'CBS thought the characters of Cagney and Lacey were 'too tough, not feminine, ... harshly women's lib... 'dykes' (cited in D'Acci: 213). Mind-sets were not flexible enough to embrace both and, instead they sustained the either/or binary only by construing the representation as a 'masculine' distortion. In post-structuralist times, however, much has been made of the collapse of binaries in language instigating the softening of sharp conceptual distinctions. In the Derridean critique of dichotomous thought-patterns, one term inheres in another. 'Maleness', that is to say, is not a discrete and essential category in stark opposition to 'femaleness' with one term ('maleness') primary, since the concept of 'maleness', in defining itself against its negatively valenced 'other' ('femaleness'), bears traces of that 'other'.

Now I should not wish to suggest that television executives and production teams have all become committed post-structuralists. But post-structural linguistics is a significant force amongst many in a field which has generally facilitated evident changes in conceptions, practices and representations of gendering since the early 1980s. It helps us to theorize shifts in mind-sets, in our ways of seeing, which may actually be effected through the gender-bending products of popular culture as experienced in everyday 1990s life. Played out in the language practices of television, then, a quite fundamental shift in the processes of constructing and reading representations in TV dramas might be discernible. Accordingly, I look to explore in what follows, the question of whether the binaries initially noted above are being challenged to the point of collapse by new representations of 'woman' and 'man', 'maleness' and 'femaleness'. A possible, but by no means certain, blurring of rigidly defined conceptual boundaries may be in process. Alternatively, where women are cast in traditionally male social roles, they may be placed on the other side of a binary divide in an inversion which leaves categories intact.

In *Silent Witness* membranes separating one category from another are much in evidence in the laboratory conditions of Sam Ryan's professional life, if not quite between her professional life and her private life. As a forensic pathologist employed by the Coroner's Office in Cambridge, England, where she is located, Sam Ryan works in hi-tech laboratory conditions.[3] In virtually every episode she is imaged undertaking an autopsy in this environment. To take a typical example, Sam is shown to be screened off by plate glass from the detectives urgently awaiting the outcome of her investigation of the suspected murder victim. Sam Ryan is dressed in a white surgical suit with a green plastic surgeon's apron over it. On her head she wears a surgeon's 'hard hat' with visor over a skull cap and nose and mouth mask. Only her eyes are visible. Her hands are gloved. She literally handles the brain and lungs of the dead man on the slab without a flinch, in marked contrast with the 'can't bear to look' reaction of the case-hardened woman detective behind the glass partition. She speaks aloud into an overhead microphone. Her voice is steady, calm, 'professional'. She is the surveyor of the male body, not a body to be surveyed by the male gaze. She is no longer there to be looked at, or is at least not simply so.

In the construction of the character through dialogue and narrative, Ryan's 'male professionalism' is stressed. Like the woman in the Volvo ad, she operates professionally in the masculine public sphere. The emphasis in *Silent Witness*, however, is less on action and narrative drive (though Sam contributes to solving cases the police are pursuing) and more on detached scientific rationalism and deduction. To draw on another illustration (a double episode transmitted BBC1, 15-16/04/98), her sacrifice in the private sphere of a stable romantic life is pointed up by the comment that she is 'wedded to her work'. Ryan does in fact have an on-going affair with a detective in the local force in Cambridge, but the casualness of the relationship is evidenced in her independence of lifestyle and reiterated in the occasional reappearances of his wife. On the one hand, this may be read as Sam being in control of her sexuality, taking what she wants when she wants it. From another perspective, however, the lack of potential for a more settled relationship might be construed as the inevitable sacrifice professional women have to make to pursue a career. A consequent need to contain the body and emotions might be inferred as we shall see below.

Indeed, in her professional life, the privileging of mind over body and emotions, is stressed in Ryan's dismissiveness of speculation and inference in favour of empirical evidence. Unlike female buddy movies or soaps in which friendships between women are given screen time and celebrated, *Silent Witness* affords Sam Ryan no women friends in whom to confide.[4] Involved, in this particular story, in a complex series of drugs-related deaths, Ryan comes to suspect the relationship between a black police woman drugs squad officer, DI Hodgkins, and members of the sub-culture under investigation. The unfolding of the narrative invites viewers to share Ryan's suspicion to the point that DI Hodgkins may actually be thought to be behind the increasing stream of murders by forcefully-injected heroin or methadone. However, even where an apparent personal interest, or intuitive hunch, leads her in one direction, Ryan readily rises above it when empirical evidence points to a contrary conclusion. In this case the murderer turns out to be a young under-cover detective, befriended by Ryan in ignorance of his dual role, who is taking revenge on the drugsworld for the death of his illegitimate daughter.

Friction between Ryan and Hodgkins is set up at the beginning of the narrative since, in pursuit of additional evidence to support a new hypothesis of the events, Ryan requests that a body be exhumed for further tests. Hodgkins opposes this move and repeatedly implies both that, should new discoveries be made, Ryan was negligent in overlooking evidence in the first place and that Ryan is merely seeking to extend her 'pet theory'. Thus the very qualities which a sceptical tradition of 'professional maleness' might impute to women, namely incompetence associated with emotional interests, are foregrounded in a conflict between two professional women. On one level, the narrative *dénouement* works to reaffirm Ryan's location of herself in the Utopian discourse of objective scientific rationalism referenced above. Ryan is frequently represented at her computer, deploying new technologies to establish the facts. The trail towards Hodgkins in the drug murders case is ultimately shown to be false only by a DNA match between samples spotted by careful observation of test results which rule her out and the under-cover detective in. Ryan's ability to admit that she is wrong when confronted by empirical evidence is marked. She is frequently self-critical when an oversight on her part has delayed the pursuit of scientific 'truth'.

The image of Sam Ryan as professional forensic pathologist is not a new trope. Jodie Foster in the Agent Starling role in *Silence of the Lambs* is perhaps the first in a modern lineage of forensic-investigative, women doctors. Indeed, there is a notable similarity between the way in which the autopsy scene described above is shot for television and that in which Starling is shot in the celebrated *post mortem* scene in the movie. The camera, looking across the body on the slab, focuses on the female pathologist as if scanning her reactions. It should not be forgotten that there is a relationship between camera and acting in the conveyance of power. In the *Silence of the Lambs* sequence, Starling defies the camera's scrutiny to emerge from subservient young woman student to independent professional investigator. Like Sam Ryan who discerns through her microscope puncture holes in the dead male's flesh, Starling proceeds to discover through close scientific investigation, the concealed, microscopic letters under the finger nails. In the process, however, she is vulnerable as a student, a trainee investigator, subject to the scrutiny of her mentors, particularly her (rather creepy) boss who seems to take pleasure both from placing her in dangerous situations and watching her survive. Clarice Starling is, of course, also thought be on the target list of Hannibal Lecter himself.

The representation of Jodie Foster/Clarice Starling serves as a benchmark against which we might read the ambivalence of the casting, performances and role construction of Amanda Burton/Sam Ryan, Gillian Anderson/Dana Scully. For each woman, though in slightly different ways, remains, I will ultimately suggest, there 'to be looked at', the recipient object of the gaze as well as the surveyor of the body. I'll return to this contention. In the interim, I turn to *The X-Files*.

Professionally, Dana Scully is also a forensic pathologist and, paralleling Sam Ryan who works for established authority (the County Coroner), she is a government agent cast initially in the role of minder of her colleague, Fox Mulder (David Duchovny). As is widely known, Mulder's crusade is to open and investigate the 'X-files', those cases allegedly suppressed by the government since they contain material too disturbing for public consumption. In particular, Mulder's committed belief in other possible worlds and paranormal phenomena leads him to be interested to the point of obsession in pursuing non-rational accounts of realms apparently beyond scientific-rational norms. Even more than *Silent Witness*, *The X-Files* is a hybrid of familiar television series genres. As we shall see, it combines the detective, the sci-fi and the romance. In terms of gendering, too, the given relationship between Mulder and Scully which motivates the series is an interesting one. Mulder is looked upon sceptically by his FBI bosses since he exhibits non-rational qualities of fancy, intuition and obsession to the point of hysteria which have typically been associated with varying shades of 'femaleness'. In contrast, Scully is constructed as the dispassionate, efficient, rational scientist who avoids emotional involvement in her work, her partner and, as I shall discuss below, with any other partner. In short, like Ryan, she performs 'professional maleness'.

Part of Scully's narrative function in *The X-Files* is to draw to closure the potential open-endedness of the paranormal stories by way of a formal report. If to be goal-oriented is 'masculine' as traditional gendering assumes (Fiske 1987), Scully deploys the scientific-rational to this end. In an early episode, for example, as Mulder lies in a hospital bed recovering from a near-death encounter with an alien hit-man, she observes in voice over:

> Many of the things I have seen have challenged my faith, and my belief, in an ordered universe. But this uncertainty has only strengthened my need to know, to understand, to apply reason to those things which seem to defy it. It was science that isolated the retro-virus Agent Mulder was exposed to. It was science that allowed us to understand its behaviour and, ultimately it was science that saved Agent Mulder's life.

When invited frequently by Mulder to accept that some of the phenomena they have encountered cannot be rationally explained, Scully is constrained by the genre conventions and by the primary construction of her persona to hold to the fundamental tenets of positivism. For, whilst aspects of the paranormal remain mysterious at the end of many episodes of *The X-Files*, the impulse to narrative closure of the detective genre conventions requires the rational Scully to place the events in an explanatory frame. Patently, however, there are tensions between this genre-affirming and gender-norm-inverting stance and her increasing personal commitment to Mulder and his history and beliefs as the series have progressed. Aspects of the romance genre are set in tension with the codes of the detective genre. Indeed, two of the key narrative hooks of *The X-Files* concern Scully's potential conversion to Mulder's point of view and her possible romantic involvement with him.

It has frequently been observed that, in ratings terms, the best narrative device a TV drama series can have is a chemically unstable mix of two attractive protagonists set up for potential, but unfulfilled, heterosexual romance. This device has become key to a number of contemporary series, perhaps indicating residual 'normative' assumptions about gender on the part of audiences, or at least TV executives' residual assumptions about audiences. Scully and Mulder afford an excellent example of such a partnership; Sam Ryan and DI Michael Connor (Nick Reding) provide a variant version.

From the outset of *The X-Files*, producer Chris Carter recognised the narrative power of keeping Scully and Mulder so preoccupied with their professional interests that only the viewers wondered if and when they would find themselves in bed together (Lowry, 1995: 16). Scully's apparently complete lack of interest in heterosexual relationships invites a range of readings from 'dyke' (as with *Cagney and Lacey*), to idealized virgin heroine unattainable or available only to the right man (i.e. in his own perception, to each of the members of the Testosterone Brigade, see below). In *Silent Witness*, the relationship between Sam and Michael is consummated but, for the reasons noted above, it is unlikely to extend beyond the casual, and thus a weaker, but nevertheless important, question remains about traditional narrative romance. As the situation is constructed, Sam is unlikely to marry and have children even though she is conventionally attractive and sexually active (I am evoking here traditional gender assumptions). It is in this aspect of the series, however, that the performance of Sam Ryan by Amanda Burton presents itself for consideration.

A decade since, Gillian Skirrow noted the relative absence of discussion of the qualities of performance in film and television which figures so much in theatre criticism. Yet, as she puts it, '[a]cting does have meanings and produces effects' (1987: 165). Amanda Burton performs Sam Ryan with an intensely private quality which suggests determined control of an inner pain. The backstory of the character of Sam Ryan is that her RUC officer father was blown up in a booby-trapped car when she was thirteen years old. The effect of this experience manifests itself, I suggest, not only in the narrative opportunities to confront death provided by Ryan's professional role but also in Burton's performance in terms of a sensitivity to hurt and untimely deaths, particularly when suffered by children. Any 'natural' readiness to take on the role of mother and family nurturer appears to have been displaced into her professional devotion. The performance by women of typically 'masculine' roles plays against the already gender-coded genres. The performances may work with or against the genre conventions.

In playing Dana Scully and Sam Ryan, Anderson and Burton both perform maleness whilst retaining a strong sense of traditional female attractiveness. Apart from the circumstances when they wear specialist gear (Scully occasionally wears a bullet-proof jacket and fatigues besides the surgical garb common to both doctors in conducting autopsies) both typically wear smartly-tailored suits with trousers in their professional lives. And, since their professional lives effectively are their screen lives, dark suits and the surgeon's dark green garb is the dominant, 'masculine' dress-code. In addition, as argued above, the two women are placed in a professional world which is encoded as male in its situation in the discourse of scientific rationalism. Whilst they find themselves from time to time in dangerous situations, however, they do not typically engage in action-adventure solutions to their plights. Both performers remain demure and largely unruffled.

Amanda Burton and Gillian Anderson are both conventionally attractive women. Indeed, through the success of *The X-Files*, 'demure 5ft 3in redhead' Anderson has become a pin-up. She has appeared on the front covers of a number of major style magazines and she has a heterosexual male fan-club of X-philes who cast themselves as the Testosterone Brigade.[5] Magazine features, however, bring out her intelligent maturity following some wild years as a teenager in Grand Rapids, Michigan (*Radio Times*, 13-19/07/99). On a smaller scale, Amanda Burton has been the subject of features which, alongside photographs emphasizing her physical attractiveness, tend otherwise to present her as a private, intelligent woman with 'a certain prim cautiousness'. Her choice of a Cherokee Jeep and her love of driving serve to hint, in contradiction of this image, at a more adventurous spirit (*Radio Times*, 17-23/02/96). In the discursive practices of the publicity machine, the personae of the actresses appear to reflect the contradictions in their dramatic counterparts, though fluid identities are perhaps ultimately anchored in a unitary discourse of the 'feminine'.

My interest, in concluding these observations, is to return to my key question at the outset in terms of a play of codes. What is happening in the language of representation, that is to say, when female actors, who remain encoded as 'feminine' and attractive to heterosexual men, are located in roles which are visually and narratively encoded 'masculine'? Does the camera look at the women as hinted in the description of the autopsy above, from the point of view of 'the male gaze' or is that gaze disrupted, even problematized, by the conflicting codes? There are several possible answers to these questions and, of course, different viewers may hold variant viewing positions.

Firstly, it may be that traditional gendering applies and that a dominant pleasure of those heterosexual male members of the Testosterone Brigade in watching *The X-Files* is in ogling Gillian Anderson. They may indeed take a particular frisson from her 'masculine-coded' image, about which psychoanalysts might have a story to tell. But, for sure, Gillian Anderson is not Pamela Anderson. That is to say, the be-suited image of Scully scarcely corresponds with the swimsuited *Baywatch* dress-code of C. J. Parker. Interestingly, however, Gillian Anderson was not the ideal casting choice of the network for the role of Scully. As she recalls, 'they wanted a big-breasted blonde'. Echoing the account of the production history of *Cagney and Lacey*, Anderson reports that, [e]very time I went in, they told me to wear tighter clothes' (*Radio Times*, 13-19/07/96: 14). This insight reminds us perhaps of the traditional heterosexual, not to say heterosexist, gender assumptions of predominantly male TV network executives, particularly in the USA. In Britain, also, as the casting of female actors such as Amanda Burton and Helen Mirren (in *Prime Suspect*) evidences, the physical attractiveness of performers in lead roles continues to be a factor in casting, even where the vehicle itself would seem to call in question traditional gender assumptions. The development of pin-up careers of stars such as Anderson, Burton and Mirren – with which they are at least presumably complicit, even if they do not actively pursue them – serves to point up, and give rein to, the residual 'feminine' attractiveness of their screen portrayals, however progressive the roles themselves may be in other respects.

A 'post-feminist' account of gendering might alternatively posit a 'both/and' way of thinking about the social roles of women such that they may be, and be represented as being, both physically attractive/sexually active and professionally efficient/successful. Similarly, men like Mulder may exhibit qualities associated traditionally with 'the feminine' whilst at the same time engaging in action-adventure. The key question is that of choice, of freedom to determine one's own life in relation to one's aspirations and desires rather than being constrained by notions of 'essential' characteristics (what it is to be a 'man' or a 'woman') or rigidly-defined role boundaries. If we consider *The X-Files* and *Silent Witness* in this light, however, there has to be some doubt about the freedom of choice. In both TV drama series, the women appear to have to choose between success in the public sphere, in a high-powered, professional domain, and a conventionally fulfilled personal or romantic life.

Scully's life might appear to be more constrained in this respect than Ryan's. Where Scully appears to have no time for the private sphere, Ryan, as noted, has an active romantic and sex life, perhaps short-term and casual by choice. What militates against the latter conclusion regarding Ryan's contentment with her lot, however, is her frequent association in the series with pain and, particularly, death. Shots of her visiting gravesides to pay her respects or observing ordinary family-life with a wistful look fuels the backstory of her own family's pain and disruption. It might not even be too fanciful to hear in the lone female voice of the series titles soundtrack (composed by John Harle) religious overtones evoking more the bride of Christ than a secular bride. At the very least, the endnote of the titles may be said to be melancholic and haunting rather than up-beat.

The final possibility I have sought to explore in this discussion, namely that a fundamental change in the language of representation may be evident in Ryan and Scully's roles, remains possible, but perhaps unlikely. A cynical view might account for the crossovers of role and genre recounted above in terms of the television industry's hunger to improve ratings. Being now aware of shifting cultural values in the light of the potentially large female audience, yet loath to lose the prime-time male audience, executives may be more disposed than in the days of *Cagney and Lacey* to entertain representational innovation. At the same time, they have learned from everyday versions of contemporary theory that ambiguity and ambivalence afford a diverse range of viewing positions. They may even have found in 'post-feminism' a means of justifying their casting of conventionally attractive, middle-class women. If crime-detection remains 'a man's genre', its structural form remains to draw the male audience, whilst the role reversal of a Sam Ryan or Dana Scully affords other meanings and pleasures. In a context of linguistic slipperiness, roles such as Scully's may be said to offer empowering pleasures to a young professional AB female audience, whilst not denying the pleasures of the male gaze to the Testosterone Brigade.

In this reading of *The X-Files* and *Silent Witness*, then, the progressive aspects of the representations of women characters remain in that suspension of conservatism and empowerment which is a feature of a post-modern dislocation of bearings resulting in undecidability. In Caughie's summary:

> [i]t gives a way of thinking identities as a play of cognition and miscognition....But,...it does not do it in that utopia of guaranteed resistance which assumes the progressiveness of naturally oppositional readers who will get it right in the end (1991:55).

For sure, the women are there to do more than be looked at; indeed they are there to be active in driving the narrative and to take on roles historically available almost exclusively to men. The representations may thus be empowering at least in the 'established terms' of a bourgeois feminist tradition. Their potential in this respect must be off-set, however, by those remnant aspects of traditional gendering and, notably, the pre-feminist constraints on the conception of a range of possible social roles for women. Where professional opportunities are shown to be available, moreover, the dilemma of the incompatibility of the public and private spheres in women's lives remains.

In narrative terms, the idea of agency driving the story may have led to a mere inversion of binaries, a practice of role reversals which, as D'Acci has observed, 'essentially accepts uncritically the basic societal structures and power arrangements' (1987: 207). Amanda Burton quite recently remarked that:

> [n]ot many [women] have the main part in a series. Most roles are token – wife, girlfriend, mother, sister, daughter – a feed to the male peacock. *I'd love to be a man for a while and see how different it is when you kick ass* (*Radio Times*, 17-23 February 1996: 18, my emphasis).

Where males once dominated the narrative, women may have taken their place in the driving seat, but the underlying issue of a 'masculine' culture of aggressive, goal-oriented individualism has not been addressed.

To take a more positive view in 'new terms', however, the frequent appearance on television screens of representations which deny easy conventional viewing positions in terms of both genre and gender may have an accumulative effect. Their aggregate influence at the psychological level of empowerment may extend beyond the domain of specific professional contexts to a sense that things are generally more open to women, and for that matter men, in contemporary life. This drip-feed impact may be as much as we might hope for from a relatively conservative medium such as television. A more radical challenge to conceptual frameworks is still inhibited by the various influences, genre and institutional, which diffuse challenge to orthodoxies. In institutional terms, the TV networks, increasingly commanded by a small number of (usually male) executives, may still tend to sustain a conservative influence by aiming to repeat the formulae for former successes. But the need to 'make the same but different', as interpreted through contemporary social changes, has brought fresh imagery and narrative play to television screens.

In sum, the binaries noted at the outset may not yet have been challenged to the point of collapse by new representations of 'woman' and 'man', 'maleness' and 'femaleness'. Indeed, in the representations discussed, it seems, at worst, that women may have been further exploited by placing them imagistically on the other side of a binary divide which, though it may have some progressive impact in 'established terms', leaves categories intact. Residual cultures, both of production and reception, are not easily shifted and radical formal experimentation, even assuming its capacities to effect such change, are unlikely in the increasingly commercial, and relatively conservative, frame of the TV screen. Furthermore, as noted at the outset, changing the mind-sets and practices of everyday life is not a simple matter of new representations. Yet, despite all these reservations, a gradual but persistent challenge to conceptual frameworks may slowly be contributing in 'new terms'.

Progressive potential, in Skirrow's view, 'is located not so much in the representations of women, but in the plot which necessarily brings performance to the foreground' (1987: 185). The placing of women in roles typically assigned to men in genres such as detective fiction which are themselves traditionally 'masculine' has this effect, as aspects of the discussion above illustrate. In addition, the hybridisation of genre contributes to the foregrounding of performance. Though its primary cause may be commercial, that chase for 'something new but different', the hybridisation of genres can itself be disorientating. The positioning of Ryan and Scully, then, at sites where the discourse of the romance, detective and sci-fi genres intersect, makes for some interesting collisions which deny an easy viewing position. The repetition of these, and similar, minor dislocations may have a cumulative impact in a medium where a gradualist approach to change is perhaps the only one possible. Whatever their functionings, they at least avoid perpetuating old habits. Ultimately, then, over time we may be dealing in a new language of representation which has gradually, and unconsciously, been informed by insights of post-structuralist theory. Ironically, the industry's fundamentally commercial pursuit of 'the same but different' might yet produce profound shifts in ways of seeing. The jury remains out.

## Notes

1   The term 'woman' is used here to indicate not actual, historical or contemporary women but constructions produced by a range of discursive practices in culture. Television is but one medium of discursive practices, though perhaps an important one in terms of the omnipresence of television today in many people's lives. The significance of 'woman' thus remains in negotiation between representations in textual constructions and other constructions circulating in culture.

2   For a review of 1970s and 1980s theoretical positions with regard to gender and dramatic form, see Dyer, 1987: 6-16. For a more recent reflection, see Mellencamp, 1998.

3   This chapter was written prior to the third series of *Silent Witness* (BBC1, Summer 1999) in which Sam Ryan has relocated to London as Professor Ryan, lecturing to pathology students at University of London.

4   Sam Ryan does have a sister who, in gender construction terms is so different from Sam that she seems almost designed to set Sam's tautness in relief.

5   Duchovny's female fan club, it should be noted for the sake of balance, is the self-styled Duchovny Estrogen Brigade.

## Bibliography

Baehr, H., *Boxed In: Women and Television*, London, Pandora Press, 1987.& Dyer, G., (eds)

Brunsdon, C., 'Men's Genres for Women' in Baehr, H., & Dyer, G., (eds).

Caughie, J., 'Playing at Being American' in Mellencamp, P. (ed).

D'Acci, J., 'The Case of Cagney and Lacey' in Baehr, H., & Dyer, G., (eds).

Dyer, G., 'Women and Television: an overview' in Baehr, H., & Dyer, G., (eds).

Fiske, J., *Television Culture: An Introduction*, London, Routledge, 1987.

Lowry, B., *The Official Guide to The X-Files*, London, Harper Collins, 1995.

Mellencamp, P., (ed), *Logics of Television: Essays in Cultural Criticism*, Bloomington, Indiana University Press, 1990.

Mellencamp, P., *A Fine Romance: Five Ages of Film Feminism*, Philadelphia, Temple University Press, 1998.

Morley, D., *Television Audiences and Cultural Studies*, London , Routledge, 1992.

Mulvey, L., 'Visual Pleasure and Narrative Cinema' *Screen* vol. 16, no.3, Autumn 1975, pp 6-18.

Nelson, R., *TV Drama in Transition*, Basingstoke, Macmillan, 1997.

Skirrow, G., 'Women/acting/power' in Baehr, H., & Dyer, G., (eds).

# Patriarchal Politics: *Our Friends in the North* and the Crisis of Masculinity

Jeremy Ridgman

## Introduction

This chapter traces the long production history of Peter Flannery's 1990s political epic *Our Friends in the North*, and explores the extent to which it shares the perspectives and techniques of the 1970s TV dramas of social comment and socialist commitment. In detailed analysis of the presentation of both context and character, it traces the potential links between a failure of political agency and a crisis in masculinity which may run in parallel with the increasing emphasis on personal, rather than political, identity in both life and TV dramas.

A drama which 'will trace four friends' sexually and emotionally charged journey from the 1960s to the present day'. With these words, a BBC continuity announcer introduced the first episode of *Our Friends in the North* on 15 January 1996. Peter Flannery's nine part series was to cover thirty years of political history from 1964, its episodes marked by the dates of key events – the miners' dispute of 1984, the stock market crash of 1987 and, of course, general elections. Yet one of the many ironies at the heart of the drama's often bleakly comic tone would be the varying levels of indifference shown towards these moments of parliamentary change. For, in an echo of the seminal political dramas of the 1970s, the Loach/Garnett/Allen quartet *Days of Hope* (11 September 1975 BBC1) and Trevor Griffiths' eleven-part series *Bill Brand* (7 June 1976 ITV Thames), *Our Friends in the North* was about the crisis of socialist commitment in Britain and the often compromised place of the parliamentary Labour Party in the struggle for social justice.

The series was also about corruption in local government, Westminster and in the police force, 'the great moral issue facing British politics' as it is described by the protagonist Nicky Hutchinson in the second episode. It drew much of its authority in this field from a close shadowing of documented events, most notably the local government housing scandals in the north-east during the 1960s involving the politician T. Dan Smith, the architect John Poulson and the Tory cabinet minister Reginald Maudling. With these and several other characters modelled on real-life figures, the degree of proximity in inverse proportion to the risk of libel, this was a saga partly readable as the televisual equivalent of a political *roman a clef*. It was also, increasingly into the second part of the series, concerned with the social fabric of contemporary Britain, its central metaphor of housing policy providing not only the interface between public politics and domestic experience along which the social realism of popular television drama so frequently operates, but also a launch pad for a rich network of associative imagery. Mary Cox's hopeless attempt to paper over the mildewed walls discovered behind the wardrobe in her two-year-old, award-winning system-built flat provides a powerful symbol of the decay that spreads beneath a social politics governed by profit and public relations. The image is echoed, in the *coup de theatre* at the end of the same episode – the discovery of the two-way mirror that lurks behind the grandiose portrait of the porn baron Benny Barratt and through which he observes and records the sexual activity of the corrupted police officer drawn into his empire. A key metaphor for the forensic project at the centre of this wide-ranging narrative is its concern with surfaces and the connections that lie beneath.

*Our Friends in the North* then may be seen as occupying what Tulloch (1990, p.37) has called the 'tight radical position' usually found in television only through authored single plays and 'the occasional authored series'. Yet, as the authored has become a less occasional form series during the 1980s and 1990s, gaining ascendancy in the schedules over the single play, so such radical television drama has become a more fluid commodity, its traditionally social-realist voice more subject to the various intertextual discourses and pleasures of popular screen narrative. The rise of the political thriller, reaching its apogee in *Edge of Darkness* (4 November 1985 BBC2, shown again in December BBC1, on 10 May 1992 BBC2 and on UK Gold on 7 May 1995 ), and the impact of Potter's *The Singing Detective* (16 November 1986 – 21 December 1986), can be seen as a seminal moment in the construction of a form characterised by the merging of the legacy of social realism with a range of generic and stylistic reference points.

It is not surprising then that an opening continuity announcement should be able to offer the intricacy and epic scope of this political drama in a conveniently neat package, combined the gratifications of sexual adventure and the heightened emotion of character interaction with a nostalgic trip through recent history. When, during the serial's protracted gestation, Michael Wearing had the task of persuading the new controller of BBC2, Michael Jackson, to keep the project going, he sold the project, according to Mark Lawson, on the appeal of the central love story between Mary and Nicky, and the retrospective on the more recent events of the Thatcher regime which a generation of 30-somethings (like Jackson) would remember as landmarks in their own young adult lives. A similar process of identification can be found in the marketing of the video release of the work. Each of the two double cassette volumes carries a publicity shot of the four friends Nicky, Mary, Tosker and Geordie. On the sleeve of the first volume, 1964 -1974, they stand together, in their early twenties, against the background of the Tyne Bridge. On the second volume cover, 1979 -1995, they occupy exactly the same positions but their costume, hairstyle and physical build now mark them out as the middle-aged characters of the final episode. In a manner typical of television publicity (Fiske, 1987, p. 120), these are clearly the characters of the drama, yet divorced from the diegetic logic of the narrative. Neither shot bears any relation to a scene from the drama and in each the actors acknowledge the camera, as they never would in the fictional context, looking out at us as if real. Post-textually, the title has come to embody a certain implicit relationship between spectator and character, the extent to which they have become *our* friends. One scene, towards the end of the final episode, presents the spectator with a correlating image but one which is firmly fixed by a very different set of looks within the frame of the realist *mise-en-scene*. The funeral of Nicky's mother, Florrie, is over and the last few guests at the wake have departed. Tosker leaves with his wife but returns, alone, to collect a child's forgotten toy – a pretext maybe. The four friends are left together in the front room and as Geordie picks out a melody on the piano their eyes meet. There are no words but as they take in each other's presence they smile as if to acknowledge the substance of this fleeting moment together after all they have been through apart. It is a scene which, in its transitory pleasure and in its binding of the spectator to the under-stated lost hopes and desires of its participants, echoes, as Bruce Dessau observed in a *Time Out* preview (1996), the gathering in the 1983 film of post-1960s mid-life crisis, *The Big Chill*.

My concern in this chapter is with the extent to which *Our Friends in the North* frames political history as a drama of domestic and sexual relations, centred on the most inti-mate and affective moments in its protagonists' lives – love, loss, death and psychological breakdown. In this respect, the narrative clearly favours the three male characters – Mary, for all her growing independence and commitment to a career in politics, is given far less actual narrative space of her own and the large majority of her scenes are predicated on the desires and dilemmas of the male partners, Tosker and Nicky. Michael Kackman, in his study of the 1950s USA television espionage drama *I Led 3 Lives*, argues that in situating its central figure, the ex-Communist and right-wing hero Herb Philbrick, as a traditionally masculine father, the programme inscribes its history of American Communism as 'a gendered struggle over the integrity of the home and the authority of its patriarch' (Kackman, 1998 p. 99). In *Our Friends in the North* also, history, the failure of socialist belief and agency and the trauma of social decay and corruption, are lived out through a gendered struggle. At its centre is the crisis of male identity, not only situated in the private life and subjectivity of the hero but played out through symbolic patterns familiar, of castration and the struggle with the Law of the Father.

'It is in the *name of the father*,' maintains Jacques Lacan, 'that we must recognise the support of the symbolic function which, from the dawn of history, has identified his person with the figure of the law' (Lacan, 1977, p. 77). The struggle with the legacies of working-class culture and socialist commitment, typically embodied in the figure of the dominant father, is a familiar one from the television work of writers such as Mercer, Potter and Griffiths. It is revisited here in the battle between Nicky and his father Felix but also developed through a number of secondary paternal relationships, some of them real (Geordie and his father, Anthony and Tosker, Felix and his own father) others more implicit such as between Geordie and his patron Benny and between Nicky and his erst-while employer and political teacher Austin Donohue. 'The male subject's aspiration to mastery and sufficiency are undermined from many directions,' writes Kaja Silverman at the beginning of her study of films made in the USA in the aftermath of World War II, 'by the Law of Language, which founds subjectivity on a void; by the castration crisis, by sexual and economic oppression; and by the traumatically unassimilable nature of certain historical events' (Silverman, 1992 p. 52). It is with the 'trauma' of British history between 1964 and 1995, with history as something that 'hurts', as Frederic Jameson, quoted by Silverman (1992, p. 55), puts it, that *Our Friends in the North* is undeniable concerned. It is a trauma lived out through the suffering of the male subject and the crisis of masculine agency.

It is worth considering at this stage the historical status of *Our Friends in the North* itself as a moment of cultural history. The series was that most peculiar phenomenon of contemporary broadcasting, a television event, 'distantly echoing,' as Nelson suggests, 'those moments when *The Wednesday Play* provoked discussion in pubs up and down Britain' (Nelson, 1997, p. 246). Such events however do not simply occur. They are to a great extent the construction of the mediated public discourse that surrounds the moment of transmission. To this extent, the success of the series – particularly as an example of the merging of progressive ideas with the high cost production values more commonly associated with the historical costume drama (which, of course *Our Friends in the North* partly was) – came to be closely associated with the identity of the BBC as a cultural institution. Within five years, we find it cited from contrasting sides of the debate about the condition and future of public service broadcasting values in a digital, post-Birtian era. 'Drama series like... *Our Friends in the North*,' declared the BBC's public report of 1999 *The BBC Beyond 2000* in confidently Reithian tones, 'have so succeeded in entering the public imagination that they have become part and parcel of our national culture' (1999, Introduction, p. 2), while Catherine Bennett, attacking John Birt for the alleged attempt in his 1999 *New Statesman* Media Lecture to present himself as the 'self-proclaimed guardian of a "golden" BBC age', points to his prioritising of 'recent jollities' such as the 1998 serialisation of *Vanity Fair* (1 November 1998 BBC1) and the popular science series *The Human Body* (3 October 1994 Channel 4) over 'more substantial achievements' such as *Our Friends in the North*.

Much of the promotion of the serial as an event centred on the scale of the project, monumentality conferring upon the text a degree of cultural authority. It was reputedly the largest commission for a drama project in the channel's history and press releases and previews spoke of the £7million production cost – half of BBC2's entire series budget in 1996 – the 40 weeks of shooting, the 160 actors, 3,000 extras and 110 locations. Above all, however, this drama of British political history 1964 -1995 came with its own production history, a fourteen year narrative that was to emerge not only as a crucial element in the publicity for the series and its claim to quality but as a reflection on recent developments in the history of the BBC and its drama output.

The lengthy and convoluted genesis of the series threw into relief its status as a public measure of the BBC's commitment to politically challenging or 'progressive' drama. Throughout this long campaign on behalf of the project, according to Flannery, the executive producer Michael Wearing was motivated by the belief that it was a 'litmus test of what the BBC is about' (Dessau, 1996, p. 19). In the *Daily Telegraph*, it was placed alongside *Boys from the Blackstuff* (10 November 1982 BBC2) and *Law and Order* (6 April 1978 BBC), welcomed it as 'the kind of thing at which the BBC used to excel in the late seventies and early eighties' (Summers, 1996, p. 17). The inner sleeve of the second volume of the video cassette even quoted a description by Christopher Eccleston (Nicky) of the series as the kind of demanding, quality drama which isn't heritage-style, theme park television, and which – certainly during the later Eighties – has been missing from British TV'.

## Patriarchal Politics: Our Friends in the North *and the Crisis of Masculinity*

At the heart of the story of the serial's development lay its origins in Flannery's stage play for the Royal Shakespeare Company, performed in 1982. Running at over three hours and with a cast of 16 playing some 70 characters, it was an epic state-of-the-nation play in the mould of Hare and Brenton's *Brassneck* (1973) and David Edgar's *Destiny* (1976), its historical sweep covering the years from the Labour General Election victory in 1964 to the impending triumph of the Thatcher-led Conservative Party in 1979. Michael Wearing who, as a producer with the BBC English Regions Drama unit in Birmingham had just completed *Boys From the Blackstuff*, saw the play in Newcastle and persuaded Flannery to turn it into a four-part serial for television. Plans for the production moved from BBC2 to BBC1, surviving various changes of personnel at the BBC until 1984, when the appointment of Michael Grade as BBC1 controller brought this first chapter to an abrupt end. Grade's populist schedule for BBC1, it was said, charged with the task of clawing back the mass audience in the ratings war with ITV, did not have space for six hours of drama about political corruption.

When Alan Yentob took over BBC2 in 1988 he recommissioned the series. Flannery's stock had risen with the success of *Blind Justice* (10 December 1982 BBC2), his five part series about a radical legal chambers. The narrative, in eight 60-minute episodes, would now cover the years 1964 to 1988. Here, however, the project ran into concerns from the BBC's lawyers, anxious about the possibility of libel writs from T. Dan Smith, Poulson and the Metropolitan Police, and Flannery gave up on the project. Interest was renewed in 1992 when the BBC Drama Department asked him if the existing scripts could be adapted to form the middle section of *Seaforth* (9 November 1994 BBC1), a northern family saga covering the years from 1930s to the 1990s. When Flannery rejected this recycling scheme, Wearing once again persuaded Yentob to recommission the entire series, now in nine parts and covering events up to 1991. After further negotiations with Yentob's successor, Michael Jackson, the series was accepted for production in 1994. Even then, the story was not over. The production lost its original director, Danny Boyle, following the success of *Shallow Grave* (released 1994, shown 9 January 1997, Channel 4) and parted company with another, Stuart Urban, over differences of stylistic interpretation three weeks into shooting. Still, also, the fear of libel writs held sway. Right up to the completion of filming, the storyline involving the businessman Alan Roe who leads a back-to-work campaign and initiates Tosker into the local Masonic lodge, was being cut down to avoid suggestions of impropriety against the multi-millionaire North-East entrepreneur Sir John Hall. Flannery even had to fight off a proposal by the legal department that the action be relocated to a fictional country called Albion, a moment wryly immortalised in the use of this name for the seedy boarding house where Geordie goes to ground.

As we might expect, the political landscape covered in the scripts that went into production in 1995 is considerably wider than that surveyed in 1982. The perspective offered from the vantage point of the dying months of the Major administration and the emergence of New Labour also had a certain piquancy, an anticipation of change seen at the time as capturing something of the zeitgeist. However, as the series neared completion, another reading of its history began to emerge. The heroic narrative of the series' fourteen-year, stop-start journey towards production now thrust Flannery forward as the subject of the story – the embattled artist, writing and rewriting, waiting, negotiating, moving in and out of other projects, his reputation rising from successes such as *Blind Justice* and the stage play *Singer* (1989). This story was to become a hallmark of the auteurist discourse that inevitably surrounded the promotion and reception of the series. More interestingly, it came to frame Flannery's own proclaimed sense of his changing identity as a writer. 'Flannery has literally aged with the characters as he has updated them', wrote Bruce Dessau (1996), while Chrissey Iley in the *Sunday Times* elicited from the writer a searching reflection on the crisis of confidence and depression he had experienced during the struggle to get the show on the road and which, by implication, had been written back into the dramatic narrative itself (Iley, 1996, p. 5). As his work on the series progressed, what had started out as a story about the politics of institutional failure had become for Flannery, a more personal project. There were elements of him, he claimed, in each of the four characters: 'Mary's Catholicism. Nicky wanting to change the world as a young man and at 50 saying "I realise that not only do I not have the answers, I don't even know what the questions are any more." I'm Geordie in my search for a family and Tosker as someone who doesn't give a fuck about politics' (Dessau, 1996). It was is the character of Nicky however that this sense of a personal identity became most strongly focused. The final episode, Flannery recounts to Sean Day-Lewis (1998, p. 187), was written a year after the rest of the series, at a time of 'acute difficulty in my private life…. In a sense I was concentrating much more on the personal stuff by the time I got to the end. You can only write about yourself, there's nothing else to write about'.

There are significant shifts of emphasis between the post-Brechtian epic performed in 1982 and the eleven hours of multi-narrative drama that finally went to air. In considering the particular discourse of gendered identity that holds the series together however, what is particularly remarkable is how the connotations of the title itself, with its oxymoronic expression of cultural unity and social divide, fluctuate with the changes in the narrative as the associations with the personal relationships between the principal characters assume precedence over the public politics of historical events. The phrase, recalls Flannery (Lawson, 1999 p2), had come from a secret memo from one oil company, Shell, to another, BP, in Africa. The oil embargoes imposed by the anti-apartheid policy of the British government were to be ignored so as to look after 'our friends in the North', i.e. the white Rhodesian government. In the stage text, the expression is used by Bourne, an oil executive charged with organising a sanctions-busting back door export of oil to Rhodesia under cover of sales through foreign subsidiary companies. Here the title clearly underlines the significance of the Rhodesia question and, by implication, the wider international dimension through which economically driven decisions in one country impact on political events in another. The further connotation of the north-south divide in Britain, evident as events in London and Newcastle become more inextricably connected, is given an even more ironic twist through this link with African politics. In a crucial scene set in the House of Commons, Bourne meets a British cabinet minister to present her with the sanctions-busting as a *fait accompli*, a trap from which the Labour government is powerless to escape without a public relations fiasco. The scene simultaneously juxtaposes this meeting with the architect Edwards' lobbying of the

*Patriarchal Politics:* Our Friends in the North *and the Crisis of Masculinity*

Conservative MP Seabrook on behalf of his Middle East business interests, and Nicky's realisation that the chain of deals upon which Donohue's housing project depends lead to an inevitable conclusion: 'I thought I was joining a crusade. I thought I was working for a future. Instead I find I'm working for a builder' (Flannery, 1982, p31).

With Geordie's recruitment as a mercenary fighting the black liberation movement in Mozambique, his meeting with a guerrilla leader and his return to London armed and determined to 'seize the power', the African political scenario develops in the stage play as a central metaphor for the interconnectedness of international capitalism. Not surprisingly, it was the emblematically named Geordie, his trajectory weaving the threads of the drama together, who could thus be read by one critic as the stage play's 'central hero' (Grant, 1982, p. 29) . By the end of the television project however, this role has shifted substantially to the figure of Nicky. Geordie's injunction to 'seize the power' becomes increasingly blighted. His journey south to the porn trade of Soho becomes one of several links in the chain that connects events in London to those in Newcastle but it is a journey of a very different kind, an odyssey of the heart and the mind, that is the basis of Nicky's re-centered role as protagonist.

The Rhodesian scenario, admits Flannery, was 'the one thing the BBC ruled out' when the series was first commissioned for television, 'not so much political censorship more a sense of unity, there being enough busting of scenes between Newcastle and London' (Day-Lewis, 1998, p. 184). We may recognise here the limits of realism and the element of ideological closure that creeps into television drama in the name of unity. What interests us here however is the displacement of the sanctions-busting storyline into the domesticated narrative of the television script. Rhodesia remains in the series only as a back-story to Tosker's employment in Episode 3 by a company exporting machine parts via South Africa. When the export is halted, Tosker is laid off with the explanation that it is because of 'the Rhodesia question' a phrase that becomes in turn his byword for the resentment that pushes him into the ideological arms of private enterprise: 'them that are in charge can do what they like and the rest of us can lump it.' In the increasingly privatised scenario of the series, Rhodesia has come to stand for the 'Other' of world politics, events beyond British shores which in their incomprehensibility force the individual subject into the laager of his own privatised self. As Rhodesia is lost from the political network of the narrative, the title's resonances centre briefly on the Newcastle-London axis – the Newcastle police officer Roy Johnson, leading the inquiry into corruption in the Metropolitan force is greeted condescendingly as one of 'our friends in the provincial police forces' – but allude more substantively to identities and personal relationships of the four central characters.

Although the four are plainly old friends in the original text there is little to link them emotionally. Tosker and Mary have a steady marriage; Nicky assists them in their housing difficulties but moves on. Indeed, the play ends with him happily married and celebrating his daughter's christening, a moment of optimism underscored by his readiness to rejoin the local Labour Party under Eddie Wells' leadership; a very different moment of liminal anticipation from the one at the end of the series as Nicky commits himself to the possibility of a reconciliation with his estranged wife Mary – 'If not now, when?' The television narrative moves between Tosker and Mary's failing marriage

(Newcastle) and Geordie's downward spiral (London) but it is Nicky who bridges these two narratives. He is the childhood sweetheart and rival for Mary's affections and the teenage friend of Geordie and his own political odyssey – from a fiery belief in the struggle for justice, his PR work for Donohue and subsequent disillusion with his entrepreneurialism, his involvement with a Trotsyist cell and his unsuccessful campaign as a Labour Militant candidate, to his eventual retreat from politics into a career as a photojournalist – is interwoven with the shifting fortunes of these two intense relationships.

If we are looking for a metaphor for Nicky's role as subject, we need look no further than his occupation as a photographer. Cinema has provided us with numerous examples, from *Rear Window* to *Blow Up*, of the photographer's camera as a scopic symbol of phallic agency and the masculine aspiration to mastery of meaning. Throughout the series, Nicky's activity as a photographer is an index of his political commitment. Photo-documentary becomes his means of engaging with social injustice, photo-surveillance of the Chief Police Commissioner's house his contribution to revolutionary activism. By the end of the series however, political photography has become his profession and acclaim for his images of revolutionary Nicaragua has brought him wealth and celebrity.

It is Mary who, in the first episode, gives Nicky his first camera, a romantic conceit perhaps but an example of understated female agency 'shor(ing) up the ruins of masculine agency' (Silverman, 1992, p. 52). 'This time,' he tells Mary after the exhibition, 'I wasn't twenty and I knew I didn't have the answer, not even the question. Only a camera. So thank you. For giving me the camera'. Photography then comes to represent the displacement of Nicky's political activism into the middle-aged symbiosis of liberal humanism and professional success but in its phallic connotations it also alludes to his struggle with political impotence as he moves from participant to observer. At the most traumatic moment for the local working-class community, at attack on a pit village during the 1984 miners' dispute by riot police mobilised from London, Nicky's camera is ripped from his hand and dashed to the ground, before he is himself caught by a blow from a truncheon, a clear example Silverman's terms, of the physical 'wound' or 'splitting' that embodies the symbolic castration of the subject (Silverman, 1992 p. 102).

The camera thus comes to embody the critical relationship between Nicky's sense of masculine identity and his political purchase on the world, his struggle with a 'lack' which is both psychological and empirical. The relationship between these two subjectivities is eventually thrown into relief in the penultimate episode. The longed -for marriage to Mary founders as her political career begins to take off, though significantly this crisis is expressed primarily in terms of Nicky's feelings of comparative futility rather than her feelings. When the relationship is at its lowest ebb, Nicky encounters Geordie, now an alcoholic sleeping rough, while photographing down-and-outs in London. His sense of guilt and self-loathing are vented in his violent sexual possession of a young photography student – a scene suggesting anal penetration and echoing an earlier shot of Barratt's brutal intercourse with his mistress. Nicky is, in turn, humiliated by the student's termination of their relationship and her rejection of his middle-aged advances.

These moments of symbolic castration associated with Nicky's political failure are echoed in the fates both of his ageing father, Felix, and of his one time ally, the Labour MP Eddie Wells. Eddie, the paragon of integrity and social conscience, fades into obscurity during the Thatcher years, his only hope of real achievement a crusade to uncover a money-for-questions racket involving lobbyists and the Tory government. This storyline, according to producer Charles Pattinson, was inserted, in the interests of Labour/Conservative balance, to compensate for the removal of a plot involving Downing Street involvement in a Saudi Arabia building contract. It focuses however far more on Eddie's stake in the enquiry than in the details of the malpractice. Like Nicky, he falls under the spell of a younger woman, the researcher who has been assigned to his office but who turns out to be working for and sleeping with the very lobbyist he is out to nail for corrupt practice. The enquiry ends ignominiously and, in a scene of apocalyptic magnificence, Eddie is felled by a heart attack on Parliament Green as the storm of October 1987 swirls around him and carries his precious papers off into the night.

If Eddie represents the parliamentary wing of traditional Labour socialism, Nicky's father Felix (augmented considerably from Billy in the stage play) is the movement's rank and file. In a drama criss-crossed with journeys, Nicky's quest for Felix's soul is the most poignant; a double-tracked pilgrimage towards the great political lesson of socialist history, the betrayal of the working class by their elected representatives, and in search of the primal relationship with the father himself. Felix's rejection of his son, particularly his disparaging of Nicky's political idealism, is expressed as a form of emasculation, rendered all the more devastating when, under the influence of his encroaching Alzheimer's Disease, he begins to wander off in search for his own dead father, eventually revisiting his own harsh treatment upon his son: 'Because he knows you're useless. You've always been useless to your family.'

At the root of this cycle of self-flagellation is Felix's experience of the Jarrow march, on which he embarked as a boy, betrayed once by his own father who stayed at home and then by the members of Parliament who sent the marchers north again with nothing to show for their journey. In search of the truth that might lay to rest the political event that has blighted the relationship between father and son, Nicky makes his own pilgrimage with Felix to a Yorkshire village where one woman remembers the marchers passing. This is similarly doomed. Her spirited account not only fails to unlock Felix's now irrevocable memory but earns from the infantile old man a seemingly calculatedly act of defecation. It is now too late and Nicky's humiliation, like Felix's before him, leaves him nowhere to go. At the wheel of the car on the way home he breaks down in tears – in control but still subject to the passive tyranny of the father – and all that is left is the dominant fiction; Felix, 'is and always has been a bastard'.

Patterns of social deprivation and injustice are also expressed through the recycling of the blighted paternal law. Geordie flees from his relationship with a drunkenly violent father and finds refuge in surrogate families, first through the patriarchal haven of Barratt's porn empire and then with a brief idyll with a young single mother and her daughter. Christopher Collins, a jobless teenager, his young mother long separated from his father, terrorises Felix and his wife, turns to a life of petty crime and is wounded in a police siege of his flat. 'Power' he shouts at his young son Sean as he brandishes his gun,

'I want my dad,' he cries as he is stretchered off. In the final episode, Sean is now a street kid, locked out from his house and 'joyriding' stolen cars. Geordie makes an unsuccessful attempt to convince Collins to take more notice of his son and is punched to the ground for his pains. Anthony, Mary and Tosker's son, derides his father's upwardly mobile affiliation to the masonic order: 'Now I know why it's called the mafia of the mediocre.' Facing a thwarted career in the police force on account of his evidence against Metropolitan officers during the strike, he ends up running case meetings to address social conditions on the local estate while his own marriage begins to crumble.

In terms of its representational attitudes to masculinity and political identities, *Our Friends in the North* locks into concerns traceable back to Trevor Griffiths' groundbreaking *Bill Brand*, broadcast by ITV in 1976. This eleven-episode narrative charted the initiation of a newly elected young radical Labour MP into Westminster politics during the second Wilson administration and his increasingly bitter and frustrating confrontation with the party establishment. The series was also marked by a commitment, typical of Griffiths' dramaturgy, to the articulation of political debate through sustained and impassioned dialogue. Brand's subjectivity however depends on a governing relationship between his political trajectory and his domestic and sexual problems. His departure for Westminster coincides with the final breakdown of his marriage, implying both the incompatibility of political career building and domestic stability and the sense that his intellectual and ideological development has been at the cost of emotional detachment from his less politicised wife. Above all, the yoking of the Brand's political development and his sexual experience is underscored by his relationship with his young feminist mistress. Their erotic relationship mirrors Brand's initiation into the more radical, extra-parliamentary politics to which she is committed, their bed becoming the site not only of sexual activity but also of ideological discussion and contemplation. As Brand becomes more and more aware of the compromises that immobilise his aspirations for fundamental political change, his crisis is expressed in terms of his own sexual impotence.

Away from the arena of parliamentary or radical activism, other political television narratives – *Boys From the Blackstuff*, *Edge of Darkness* and *GBH* (6 June 1991 Channel 4) for example – have also depended in varying ways on the relationship between the vulnerabilities of masculine identity and the crisis of agency. Paula Milne's *The Politician's Wife* (16 May 1995 Channel 4), transmitted a matter of months before *Our Friends in the North*, begins to redress the balance with its exploration of the female subjectivity of its protagonist Flora Matlock, out to avenge the adultery of her Tory cabinet minister husband. Milne's particular processes of representation however – a full-blooded embracing of melodrama and an objectification of Flora as a dyed in the wool Conservative herself, capable of as much treachery as her husband – allow for a more critical separating out of her sexual and political identities.

It may be regarded as a function of television drama's continued submission to the aesthetic of social realism that social issues are 'personalized rather than located in broader social or political processes' (Nelson, 1997 p. 117). Certainly there is nothing in any of this work to approximate the more avant-garde form of *Heimat* (German Federal Republic 16 September 1984, shown 9 April 1986 and 22 Jan 1993 by BBC2), a series which in terms of theme and historical scope has similarities with *Our Friends in the North*. Flannery's epic serial however is the closest British television has produced to the scale and social complexity of the nineteenth-century novel so admired by Lukacs (1962) for its capacity for following typical characters through the vast sweep of socio-historical movement. With its often seamless shifts in style and tone, from romantic melodrama to crime thriller, from comedy to bleak realism, it is a far more open text than might at first appear. Through the gaps, we can see a deeper politics at work, not merely the political expressed as personal but the 'libidinal politics' (Silverman, 1992 p.3) underlying the masculinised experience of historical trauma.

## Bibliography

Day-Lewis, S., *Talk of Drama: Views of the Television Dramatist Now and Then*, Luton, University of Luton Press/John Libbey Media, 1998.

Dessau, B., 'Tyneside Story', *Time Out*, 10-7 Jan. 1996, pp. 18-19.

Fiske, J., *Television Culture*, London, Methuen, 1987.

Flannery, P., *Our Friends in the North* (RSC Pit Playtexts), London, Methuen, 1982.

Grant, S., Review, *Plays and Players*, No. 347 (Aug. 1982) p. 29.

Hellen, N. 'BBC cuts over legal fears', *Sunday Times*, 3 Dec. 1996, p. 10.

Iley, Chrissie, 'The history man with more than a touch of class', *Sunday Times*, 28 Jan. 1996, p.5 (XI).

Kackman, M., 'Citizen, Communist, Counterspy: *I Led 3 Lives* and Television's Masculine Agent of History', *Cinema Journal*, 38:1 (1998), pp. 98-114.

Lacan, J. (trans. Sheridan, A.), *Ecrits: A Selection*, London, Routledge, London. 1977.

Lawson. M., 'Friendly Fire', *Guardian*, 1 Jan. 1999, pp. 2-3.

Lukacs, Georg, *The Historical Novel*, trans. H & S Mitchell, Penguin, Britian, 1962.

Nelson, R., *TV Drama in Transition: Forms, Values and Cultural Change*, Basingstoke, Macmillan, 1997.

Poole, M. and J. Wyver, *Powerplays: Trevor Griffiths in Television*, London, BFI, 1984.

Silverman, K., *Male Subjectivity at the Margins*, New York & London, Routledge, 1992.

Summers, S., 'Why has this serial taken fourteen years to reach our screens?' *Daily Telegraph*, 9 Jan. 1996, p. 17

Tulloch, J., *Television Drama: Agency, Audience and Myth*, London, Routledge, 1990

# Ga(y)zing at Soap: Representation and Reading – Queering Soap Opera

## Stephen Farrier

## Introduction

This chapter generates a vision of queerness and soap opera, emphasising both the pleasures and the position of the viewer. More specifically it explores the way that queer may be applied to soap opera. Starting with an investigation of what constitutes soap opera and how it may best be described, the chapter focuses on bardic function and its relationship to the viewer. Moving on to examine what queer theory/method is, the chapter then proposes a working definition which, through the use of a fragment of communication theory, is then referred to a selection of soaps to see both what soap does, and how queer looking can sometimes reveal disavowed desire embedded in the narrative. The chapter concludes with a description of the usefulness of queer looking and the question of its relevance to mainstream texts.

Queerness is a lot like soap opera. It can be fun, rambunctious, hilarious, touching, exciting and can sometimes be downright irritating and annoying. However the soap is seen as central to the television experience of the last decade. Newspapers are fascinated with the characters and the actors who play them, often featuring them on the front page to entice the public to buy a publication. There are documentaries about how soaps are made, about how fans can become utterly obsessed by the programme, how they are written, how they come to be on our screens, how storylines are sometimes leaked to the press, or are jealously guarded (and even printed on non-copyable paper). The popularity of soap is secured in the top ten ratings consistently. What is all the fuss about?

Soap is instantly recognisable in the many guises it has. Usually it is situated within the domestic domain and is concerned, primarily, with interpersonal relationships. However, attempting to define what constitutes soap by its contents is perhaps too large a task to be addressed here and to decipher what its contents might mean is an even greater task and its fruitfulness is questionable, as David Buckingham (1987) comments:

> Ultimately, then, the text does not 'contain' a meaning which can simply be extracted and defined. Yet if we cannot say what the text means, we can at least begin to describe how it works – that is, how it enables viewers to produce meanings. (p.203)

If we accept Buckingham's premise then soap can be at least described in a structural sense, by looking at how it works on a level other than narrative, and I would say it does some very specific things. Soap encourages its viewers to become involved with the programme by the use of structural devices including cliff hangers, snares and by placing the viewer in a 'knowing position' where they are privy to the motivation of a character:

> viewers are often able to make distinctions between a character's outward appearance – that is how they present themselves to other characters – and the underlying essence which constitute their 'real' feelings: here again, the viewer is placed in a position of knowledge. Buckingham (1987. p.77)

It is these 'predictive pleasures' that provide many points of entry for the viewer. It encourages the viewer to become involved with the narrative, to second-guess actions of characters, and to place in the living room (or wherever the television is watched) issues for the viewer to become engaged with. In this way the soap resides in the public arena where it invites, questions, complicates, and comments on the culture in which we live at the moment of the soap. It is often topical and through its porosity asks the viewer to comment. These 'gaps' (which I will come back to later) or 'invitations' to become involved with the soap are crucial to this chapter, as is the notion that viewers are 'readers'.

It is with the viewer this chapter is concerned, and principally the queer viewer. Starting with a description of television's bardic function the chapter looks at what queer is and how it may fit in with some of the ideas generated by bardic function. By generating working definitions of what constitutes queer, this chapter seeks to discover what strategies are employed by a queer reading practice to enter into the soap. Queers can be said to be situated on the fringes of the viewing public, yet soap affords to them an arena for playfulness, speculation and representation. The 'nature' of soap, I argue, through its bardic function, drags the representation of queers to the centre of culture – a place where they do not fit. In doing so the queer then rips at the fabric of the soap, exposing those places which are usually kept under wraps by the soap. Queer readings seek out, and to some extent generate, these ruptures allowing those people who call themselves queer to connect with a mainstream programme that may not be available for them in other televisual forms.

Through the use of a fragment of communication theory, I generate a queer reading practice lexicography that can be used to explore the experience and processes of queer readings. Linking this to an analysis of the structure of soap opera I then go on to explore in what ways the seams, and semes, of soap fray under the nipping of a queer viewing, whilst also attempting to reconcile the concept of representing queers in soap.

## Getting Down To It: Bardic Function, Socio-Centrality and The Ex-Nominated

As a prefix to looking at television soaps, it is useful to define what television is in terms of its effects on the people who watch it. I am not talking here of the ability of television to change people's behaviour, I refer to what television functions as in society. Fiske and Hartley (1978) look upon television as a cohering factor in our lives:

# Frames and Fictions on Television

> Television functions as social ritual, overriding individual distinctions, in which our culture engages in order to communicate with its collective self. (p.85)

It is this social ritual that they choose to call the bardic function of television. The function of television as a bard, reflecting back to society what it likes to think about itself, is a crucial aspect of television's functions.

Fiske and Hartley (pp. 85-100. 1978) also explain why television is 'bardic'. They specify bardic because a bard functions as a mediator of language; structures messages to the needs of the culture; the bardic mediator occupies the centre of its culture; it is an oral, not literate voice; it is positive and dynamic; the bard is involved in 'clawback', bringing the message back to human socio-centrality. Clawback is of most interest here as it refers to the humanness of all things. Humanness clawback can be seen clearly in natural history programmes where anthropomorphism is used a great deal. Clawing back to the socio-central position of the human in society presents many problems for humans that do not live in that position of socio-centrality. In real terms, the process of clawback takes a character that may not occupy a central position in society, displaces them into a position of socio-centrality, and then naturalises that position. In this way individual distinctions are glossed over and representing the non-socio-central becomes a matter of centralising them. This clawback be seen at work in almost all soap operas. The 'madness' of Joe in *EastEnders* (a soap opera shown from 19 January 1985 on BBC1) is a good example of clawback at work. Although Joe acts strangely (and the viewer is aware of his illness) no one else in Albert Square recognises this for some time. This allows the viewer to look at Joe as different, but also the same. He is ill and through his illness he is clawed back so that the programme can 'educate' us – we are Joe, it could be us. When Joe's humanity is taken away from him and we see him in a hospital dribbling and lifeless he is still clawed back through the sympathetic gaze of the camera that encourages us to look upon his illness as distinctly his, but also to play up the 'human tragedy' of it. Joe at this point in the soap is seen as one of those people on the margins, the mad, the incarcerated. Yet through the narrative structure of the soap we are allowed to see his humanity, as long as his reality is played off against the central reality that the rest of the square share. His difference is shown, but his reality can only be hinted at through the codes that sign his illness – codes that are directly related to the socio-central. Through signing these codes early enough for the viewer to see what is going on before the other characters (who must be extraordinarily insensitive to them) do, Joe's illness becomes marked by the codes of the central, not by the codes that, perhaps, someone in his marginalised position may use. He is clawed back by codification through the 'like us-ness' of the character, not through other codes that may explore his condition and ultimately lead to a challenge of socio-central thinking about mental illness. (This has particular ramifications when we look at how the socio-central signs sexuality through its codes, see below.)

Fiske and Hartley (p. 176. 1978) use the term ex-nomination in the Barthian sense to express the way that television ex-nominates people like Joe who do not occupy the central ground of culture. Fiske and Hartley through their discussion of *Ironside* show that when the ex-nominated are represented their relationships with those who are socio-central become naturalised (that is, their foundation is taken to be a position of centrality). The issues involved in the formation of their relationships are not 'fore-grounded for inspection or criticism, but appear[s] as the natural order, and as such does not require any conscious statement' (p.176). In terms of the instance above, we are not fully offered an explanation of Joe's illness, it just *is*.

The ex-nominated do not occupy the central ground of culture and as such cannot find representations of themselves in programmes (other than those who are natu-ralised). In order for the bardic element of television to communicate successfully with culture, according to Fiske and Hartley (pp.88-89. 1979), it must articulate, implicate, celebrate, assure, expose, convince and transmit its ideas. These ideas are then negotiat-ed by the response of the viewer according to their own self, circumstances and experience. Fundamental to this negotiated response is the viewer's relative positioning to socio-centrality, which identifies them with 'dominant ideologies'.

Yet the ex-nominated reader may find it difficult to negotiate an acceptable response (acceptable for the subject) because they cannot connect themselves with a socio-central position. However, soap operas such as *EastEnders, Coronation Street*, and *Brookside*, attempt to place the ex-nominated within the fabric of their programmes. Giving voice to the ex-nominated character has problems for the soap's ability to occupy the central ground of society, simply because of the nature of ex-nomination. Ex-nominated people can be said to be invisible. *EastEnders* and other soaps attempt to circumvent this slip-page by means of clawing back the ex-nominated to the position of socio-centrality. That is, they emphasise the 'like us-ness' of the ex-nominated character (where 'us' is the sum of the socio-central).

The first gay kiss on *EastEnders,* although a mere peck on the cheek, caused quite a stir and although there have been more 'full on' gay kissing since, the Colin and Barry kiss serves as an ideal example of making the invisible visible. Up until the point of the kiss the characters had hardly ever touched – not the makings of a great relationship. Yet when they did kiss it was in Colin's flat, a safe space away from the children (inter-diegetic children, that is) and the love they shared was exposed in a way unseen in British soaps up until that point. Yet the codification of their surroundings and the language with which they spoke of their relationship was coded in terms of the norma-tive semantic stock of the socio-central. They could have been a heterosexual couple. The codification of Colin's flat, and the codification of Colin and Barry was just as all the other characters in the soap. They did not use any 'gay-isms' nor did they significate in a queer way. They were presented to the viewer as the same but different. As if gay rela-tionships were constructed and maintained in the same fashion as the other heterosexual relationships in the programme (I am working here from an assumption that in a hetero-sexual relationship there are two different sexualities at work, and in a gay relationship there are two of the same sexualities present and that this must have ramifications on the sort of signification the types of relationships would 'deal' in).

## Coming to terms

I said above, that queer is like soap; I meant this in a very literal sense. Queer can be tricky to get hold of and defining it can be immensely difficult. Queer can be seen like a bar of soap in a bath – in an attempt to grab, it slips and falls back into the bath, in doing so it muddies the water a bit. Any subsequent attempts to grab it also succeeds in a slip out of the hand and the water is further clouded. In a sense the notion of queer is always just out of reach; the more it is grabbed the more slippery it becomes, the more difficult it is to find again in the increasing opaque water. Yet each time it is grasped, even if only momentarily, it leaves a residue on the hand. This is the experience of queer – something that is there, but not easily got at, a shadow in the water, a sensation on the skin.

In December of 1991 in a club in Manchester fliers where distributed which read:

> **Queer** is not about who you fuck.
> **Queer** is not about how you fuck.
> **Queer** is not about when you fuck.
> **Queer** is not about if you fuck.
> **Queer is what you fuck:**
> > Fuck boundaries.
> > Fuck gender.
> > Fuck the lesbian and gay community.
> > Fuck labels (Anon. c.1991).

The quotation has many uses. First of all it is an early definition for queer on this side of the Atlantic ocean. Secondly it shows that queer when it first came to these shores was a rambunctious radical idea (even the use of the term 'queer' was debated at length and still is). Thirdly the quotation gives this chapter a point of origin that resides in the public arena, like soap opera does. Rather like soap, the quotation asks its readers to become involved in the issues it raises – it invites the reader to question (although I would perhaps say it is not aimed at those people who could be considered as occupying the central ground of culture). Existing within the public arena is a very important part of where queer works. This is vital because as queer as a concept becomes investigated by academics it loses its radical edge as it is formulated within more 'accepted' terms and disciplines – it is taken out of the hands of 'ordinary' queer people (if such people exist, of course) and is present only as high theory.

Definitions of queer are diverse and abundant. In fact to attempt to rank, order and group them would be a task too huge for this chapter (and also against the 'nature' of queer). Yet they do have many commonalities. They question such normal formations as gender and sex (that is how they are constructed rather than a 'given reality') and they also question what it is to call oneself a homosexual, a lesbian, a gay man, a transsexual and, indeed, a heterosexual. The following definitions are all involved in the call for validation, for other ways of thinking, for other ways of being, and importantly for this chapter, for other ways of seeing and reading soap opera.

Queer theory/practice/methodology has no fixed meaning, no dogma, no seminal work which remains constantly 'citeable'. It has been defined as many different things in critical theory. Halperin (1995) refers to queer as

> An empty placeholder for an identity that has yet to be fully realised... an identity in a state of becoming rather than as a referent for an existing form of life. (pp. 112-113)

# Ga(y)zing at Soap: Representation and Reading – Queering Soap Opera

Eve Sedgewick (1994) expresses queer as:

> ...the open mesh of possibilities, gaps, overlaps, dissonances and resonances, lapses and excesses of meaning when the constituent elements of any one's sexuality aren't made (or *can't* be made) to signify monolithically. (p.8)

Chris Woods (1995 p.29) describes it as 'iconoclastic' and 'anti-ideological' (sic). Burston & Richardson (1995) take many definitions of queer and encapsulate what it is queer does:

> Queer readings of culture can take many shapes, drawing on as they do a wide range of disciplines. [...] What they all share is the understanding that cultural texts do not have single meanings, that what is denied at the level of narrative can often be deciphered through clearer inspection of cultural codes. (p.120)

In a more formal sense Tasmin Spargo (1999) describes the relationship queer has with theory:

> Queer theorists' disenchantment with some aspects of gay and lesbian politics is not simply a rejection of the normativity of those particular categories, but rather derives from a different understanding of identity and power. If queer culture has reclaimed 'queer' as an adjective that contrasts with the relative respectability of 'gay' and 'lesbian', then queer theory could be seen as mobilising queer as a verb that unsettles assumptions about sexed and sexual being and doing. In theory, queer is perpetually at odds with the normal, the norm, whether that is dominant heterosexuality or gay/lesbian identity. It is definitively eccentric, ab-normal. (p.40)

Woods (1995) flags up the notion that queerness as an identity is for some people a process of 'identification by commodification'. There is, according to these people, a 'right way to be in your face'. Queerness as a grass roots definition has turned from a political slant to a 'type' of queer (the politico queen) that fits into other 'types' (the clone, acid queen, ice queen etc.). In effect queer at this level has become hegemonised by the dominant gay identity and all that goes along with being a 'real gay'. That is queer serves as a 'short-cut' like the 'stock characters' that soap often employs such as the 'busybody', the 'tear-away', and the 'dark horse' as Geraghty (1991) comments:

> It is in the nature of the soaps' ideal community that it can draw into itself all sorts of characters – the grumpy, the cantankerous, the complaining, the eccentric – and that they do not need to be transformed into ideal types for the community itself to be celebrated. (p.86)

I would argue, however, that soap does need to bring those people that perhaps are 'undesirable', if they are to be seen as sympathetic, towards socio-centrality in order for the community to be celebrated. Rather like the relationship between Colin and Barry, the emphasis within the programme was placed on the similarities between the gay rela-

tionship and the straight relationships in the programme (which were drawn as norma-
tive). Looking at representations of gay characters is a good way to look at clawback in
action.

Yet, it is worth remembering that 'queer' is not a synonym for 'gay'. Queer as a
concept has grown from notions of gayness, but has changed to include all of those
people who are marginalised in some way – those whose sexuality doesn't 'work' within
the gender imperative. So, in fact, queerness can encompass those people who call themselves 'straight'.

Queer, theoretically, addresses the differences of people and welcomes people who
may not fit well into the category of gay. Queer inhabits the gaps created by monolithic
binarist views of the world. Sedgewick expresses queer much as this chapter sees it;
queer gives marginalised people who will not or cannot fit into the gay category, a place
to orient themselves in the context of their lived experiences. There are problems for
some with this definition as it sees the collapse of the notion of gayness, of homosexuality. As queer invites all those who are marginalised by society into its arms the nature of
queerness changes from a place inhabited by non-heterosexuals to a collection of the
marginalised. So those who may have been attracted to the nation of queer through its
gayness (but with an attitude difference where all that is gay is not necessarily good)
may become invisible once again as homosexuals. This invisibility has ramifications for
soap, as soap, as I have mentioned above, seeks to expose that which is unseen (albeit
within strict clawback preconditions). Here is an apparent paradox, queer theory is
littered with them, queer looking can make 'invisible' the viewer in terms of their identity, yet it is this very identity that uses a reading practice that seeks to expose that which
is unseen in the soap's representations.

Taking the lead from those whose bodies or sexualities are outside of the dominant
significations, queer looks at the way identities are built within the social/cultural mix
of their particular place. This is most easily seen in the idea behind the reclamation of the
word 'queer' as a tag for identity, where naming oneself is seen as a liberating effect
(especially as a queer, ironically, is by definition indefinable). Playing with identity,
uncovering that which is unseen by allowing identities that would not normal have
signification, to significate, is an effect of queer. It attempts to give voice to those identities and bodies that have not had signification outside the strictures of socio-culturally
validated means of existence.

Soap opera also attempt to give voice to those in culture who perhaps haven't had
one in such a way before. I am thinking here of the way that *EastEnders* gives central plot
lines to gays, lesbians, single mothers, criminals, disabled people, and all manner of
marginalised people. Also *Coronation Street* has featured transsexuality which can be
seen as a surprising and radical addition to an established programme.

For queers in soap the ideas of bardic function, socio-centrality, clawback and ex-
nomination can be looked at in two ways. Firstly, the theories can be looked at from a
queer perspective themselves, generating questions about how people who don't occupy
the centre ground of culture manage to watch soaps. Secondly, the theories seek to place
the viewer in a position that the queer does not, can not (and I suspect) will not, fit. Both
these ideas about the theory beg the question how does the queer reader, read?

## What does a queer look like then?

Queer readers necessitate a different type of looking at texts, and queer characters in soap often generate problems of 'authenticity' and 'representation'. The queer characters in *EastEnders*, *Brookside* and *Coronation Street* all have 'issues' of representation attached to them. Simon and Tony, two gay lovers in *EastEnders*, can be seen as the two straight-est-looking gay boys in London, their representation doesn't look convincing to a gay audience. I would argue that this is because the gayness of the characters has been 'clawed-back' and that the readers of the image who see them as unconvincing use a queer reading practice. The two gay characters are signified within codes of recognition that are available to the central reader. To the queer reader they can be seen clearly as a rather dubious construction – they are a carbon copy of a 'straight' relationship where only the object of desire is different, rather than a relationship that signs itself as differ-ent. To a queer reader the gay relationship in this instance says nothing other to them than a representation of a straight relationship, what it does show is the implicit constructedness of all of the relationships present in the soap's fabric. Queer looking will often 'see' things that perhaps aren't available to the reader who is not using a queer reading practice. This reading is facilitated by the reader having a different orientation towards a text than a reader who is socio-central.

Communication theory offers a model that will help describe this queer reading. The minimal ABX system as described by Gerbner in 1953 (see Figure 1) sees communication as something that happens between A and B but with reference to X, the social factor inherent in communication. If the X factor was a formal situation, the way A would talk to B would be dependent upon that formality. Similarly A would talk to B differently if the X factor was informal (even then there are ranges of informal situations such as an informal interview and meeting friends). In the minimal ABX system the responses that A would give to B, and B would give to A would fluctuate dependent on the changing X factor. In queer terms if X is removed, that is the social factor is ill-defined or undefined, A's conversation with B, becomes very difficult because the rules of communication are shifting in unsuspected or new ways.

Minimal ABX systems then are in a constant state of movement. If we figure X here to be socio-centrality, or the social context of looking (if the looker occupies the centre ground of socio-centrality) and the relationship between A and B to be the reader (A) looking at the soap (B), then the system will fit differently for different readers. The X factor can never be fixed because the relationship with soap is different from one reader to the next dependent upon how far they are from socio-centrality. Also as the soap plays out its representations the relationship between A and B will change along with the reading of A. When in a constant state of flux the relationship between A and B can be difficult. One example of this is within the early days of the queer movement. A gay person taking on the label queer results in a reversal of effects. The X factor becomes difficult to grasp. In almost all situations calling someone queer was an insult, yet now those very people to whom the word was offensive wear tee-shirts declaring that word. What is going on? Reversals of effects such as these stretch the minimal ABX system to breaking point. How can queers be insulted when they call themselves queer? Such questions are rooted in the notion that the basis for communication (or insult) has

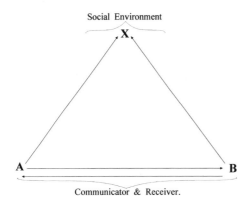

Social Environment

**X**

**A** ————————————→ **B**

Communicator & Receiver.

*Figure 1. The minimal ABX system. If A changes, B&X will change also. The system always has a need for equilibrium and is dependant upon A&B's relationship (positive or negative) and their mutual relationship to X.*

shifted, the X factor in the ABX system has revolted.

In terms of looking the minimal ABX system can create some interesting ideas about the context of looking. If we take A to be the viewer and B to be soap and X as the context of looking (the socio-centrality of A), the system rests on a notion that looking is not only situated within the viewer (A) and the soap (B), but also in the reading A has of the soap. It follows that looking not only depends on subjectivity, (how A views the representation) but also on the complex interface between subjectivity, the soap and the representation and social context of the expression of subjectivity. For instance a lesbian watching the relationship between Beth and her lovers in *Brookside*, may perhaps notice that the representation is unconvincing – that the relationships she has are drawn from a different perspective. The X factor in looking is perhaps in a different position for a queer reader than for a non-queer reader. Queer readers have a more mobile X factor that is necessitated by the relationship between their reading practice and the trajectory of the soap. Soap, as it has been said, contains the sort of structure that is full of gaps that allow the viewer to enter into it in many different ways and at many different times. Also Sedgewick (1994) describes queer as inhabiting these gaps. Queer's X factor, then, is engaged in making a different type of text than, perhaps, a normative reader might, by inhabiting these gaps and generating a new text.

The X factor in looking is perhaps best described by metaphor. Social factors can, metaphorically, be seen as an orientation to a text. For instance it could then be said that the white masculine male mainstream heterosexual looker (the socio-central subject) situates himself directly in front of the soap. Others may place themselves in different reading positions, so they may stand sidelong to the text. The painting *The Ambassadors* (1533) by Holbein displays this anamorphosis, an object that has to be seen from a certain angle to be recognised. The Holbein painting shows two men standing amongst symbolic bric-a-brac of the sixteenth century. On the rug beneath their feet is a grey smudge of paint. Yet if the looker stands sidelong to the painting, the grey smudge becomes a skull. The looker has to skew the look, or change the X factor in order to see the skull.

In the Holbein painting, the viewer has to look at the text from a non-central position in order to see the skull. The direct onlooker does not see anything but a sort of grey smudge. The skull does not interfere with the 'main' image. The painting can serve not only to examine the X factor in relation to looking, but also as a justification for queer readings of soaps. Queer readings are often placed (patronisingly) within the vague categories of 'marginal' ideas, or the ever ready set of anomalous readings entitled 'against the grain'. Yet queer readings can be as sharp as Holbein's skull. What queer readings do not share with mainstream readings is the orientation towards the text, the same response to the X factor, a different position on the fringes of socio-centrality.

Queer readings can be seen as a different response to the X factor. So the communication between A and B becomes distorted because the system is in a state of dis-equilibrium. That is A and B have differing responses to the X factor. In the same way texts that have been called (or can be called) 'queer' disrupt the normalised orientation towards the text (i.e. front and centre). In fact queer texts are anamorphotic, or at least require the looker to walk around the text to get a clearer image (of one is available at all).

So the nature of looking, or the type of looking that is available to the viewer is predicated on the type of media the text is presented through. Along with this comes the notions of codes and conventions of the particular texts that exist within culture (which are themselves in a state of flux over time). Queer readings appear to be very dense to the central position and involve a certain amount of anamorphosis. Queer readings of mainstream soaps often do not interrupt the ABX system in the same way that the queer text does, but they do leave a smear on the text, like the skull on the carpet.

For instance a queer reading may look with desire when any desire is disavowed within the soap opera itself. If we look back at the representations of Joe's 'madness' we can see the image of him shirtless, writing the word 'evil' on his chest, is ripe for discussions of desire and 'madness'. Yet Joe at this stage in the drama was 'madness', he wasn't a mentally ill person, he was mental illness. He had no sexuality, yet the image of him half-naked allowed a sexualised version of the image to be read by those who desired to read it as such. In the soap itself the image was presented as disturbing and sympathetic, to other readers it was downright sexy (hence the endless replications of it in the gay press and the deluge of fan internet sites that sprung up around the character, see Pattenden 1998 & McCarthy 1999). Yet this image is open for many versions of queer looking.

If we take one of the basic ideas of queer theory that states that identity and sexuality are 'made' by discourse and not something within us, but mapped across us then we can look at the image of Joe from this perspective and question in which ways the image is both a result of and a generation of identities. Joe lives within the social world of Albert Square, this is where we can see his identity being made. He lives within the discourse of the socio-central, and as such he does not have a way to significate himself. We don't see any of the soap from his point of view, we see it from what appears to be a natural and unmotivated point of view. In fact we see his 'madness' from a point of view of what it is to be 'normal'. The only way that Joe appears to be able to significate himself is by writing on himself – he marks his body with its difference (and it is rather interesting to note that a lot of queer performers also write on themselves – they mark their difference in a textual and readable way, see Keith Hennessey performing *Saliva* in Mohr (1992 p.207)). The combination of the desire for his body and the marking of it as 'other' (just in case any of the viewers were not familiar with the code of covering a room in tin foil as an index of mental illness, Joe writes it for all to see) we can see an interstices through which a queer look can penetrate and create a new version of the text. Perhaps a queer reader would identify (with) the image in terms of its explicit nature, or that he is codified as an outsider – that the soap can never fully explore or allow its viewers to understand him.

Reading soap opera, from a queer angle requires the reader to constantly vacillate between the position of socio-centrality and a marginalised position. This fluidity in looking comes with a queer reading practice. One way of understanding this concept is to look at representations from other cultures and historical periods. As the notion of reading changes so does the minimal ABX system. When we look at historical or 'other' cultural representations from front and centre, they can make very little 'sense'. I would say that this is because the representations are subject centred, they represent the relationship to X the socio-centrality of cultures to which we do not belong (or cannot in the case of cultures that were before our 'time'), as Brennan (1996) comments:

> The idea that we see the same way at all historical points is of course untenable. Even so, we persist in thinking that, say, the Egyptians painted the eye the way they did because they were a bit thick, or the medievalists showed perspective the way they did because they were a bit slow. We do not credit them with their own critiques of a subject centred perspective, even though we are quick to spot, say, cubism's relation to subjectivity. (p.227)

If we accept the concept of queer reading and its vacillation, the production of cultural artefacts will always be a result of the conceptions of the subject at that specific time (and soap issues are often products of their time). It could also be said that any critique of looking is also as a result of the social factor X. In this way the conception of queer looking, or moving from front and centre to a sidelong glance is a result of the reading practice of queer people at this time in the late twentieth and early twenty-first century.

Using an ABX system as an exploration of queer readings could just as well be applied to other sorts of programmes, paintings or theatre. Soap's structure, however, provides an ideal and fertile ground for the exploration of a queer reading practice, through its invitations to, and tacit acknowledgement of, the queer reader.

## Getting it Straight

Soap's structure makes use of techniques that invite the viewer to become wrapped in the narrative. The use of cliff hangers, snares and other hermeneutic devices the structure of soap asks its viewer to become engaged in the narrative by inviting them to predict and discuss the lives of the characters. Predictive pleasures allow room for the viewer to enter into the story of the soap and become part of the programme. Implicitly leaving room for an audience within the soap allows for the viewer to enter the interstices within the narrative; the soap tacitly invites the viewer to generate their own text. Adding these predictive pleasures to the manifold story lines available to the viewer, the viewer has many entry points into the soap. Also, these structural techniques and polysemeia broadcast that the queer reader is welcome into the text – that other readings are encouraged. Even if the soap, perhaps, does not want to court queer readers, there are many points of entry that queer readings can infiltrate.

Queer readings can take a sidelong glance at the characters, their situations and their community and generate their own ways to contact, or connect, with the programme. By fine tuning a queer reading practice, those who are not placed in socio-centrality can enter the narrative in a similar way to those the soap is aimed at. Yet the queer reader perhaps does this with a knowledge of the other ways of looking at the soap, and perhaps performing this looking simultaneously. By tuning into the way scenes are produced a queer reading can glimpse the fiction in a way not available to more main-stream central readings. That is, the queer reader can see the constructedness of all of the soap (perhaps something that soap's realism works to disguise). Queer readers see.

By viewing a queer character, which necessarily has to be clawed-back from the ex-nominated to socio-centrality to emphasise the 'like us-ness' of the character, queer readers can see how that character is represented. No amount of clawback would truly allow queer characters to exist in the same way as the mainstream characters do, but to the front and centre reader the representation is transparent – it needs no explanation. To be a queer watching a queer representation, however, can be quite different.

If we are to accept the bardic function of television then the representation of queer within soaps would be impossible. Queer, itself, is outside of the centre ground of culture – it is hegemonised, it is no longer queer, it becomes more slippery and runs just ahead of the hegemonising hands of socio-centrality. Yet to represent queers in a soap the queer has to be clawed back, hence the queer is no longer a queer.

Reading practices vary, however, and I would say that the majority of readers are not queer readers, they are situated in the centre, and it is worth mentioning that some of these mainstream readers are gay and lesbian, just as some queer readers are 'straight'. Soap then comes at the mainstream viewer from front and centre – they look from a place directly in front of the soap. It hits them in the face, as it were. The queer reader sits to the side of the soap and as it leaves the screen it flies past and they can see the constructedness it seeks to hide. As when looking at a subtle bas-relief, the best way to look at the construction of the picture is to look from the side – to see the contours and shape of the front of the picture whilst not standing in front of it (where these details might not be available to the looker).

Queer reader's ABX systems then are in constant mobile action, they flicker from the front to the side, around the back and over the top of soap. There is great pleasure in this movement and playfulness in reading. It is perhaps this mobility that allows queer char-acters to function within the narrative of the soap for a queer reader. The queer reader, regardless of the representation of a queer within the soap or not, is attracted to the constructedness of the soap – glancing the underskirts which usually remain hidden. This type of reader can see the constructedness of all characters and situations in soap in a way that is not easily accessed by the socio-central reader. A queer represented in a soap to a front and centre reader is also acceptable (within limits) because through soap's bardic function the 'like us-ness' of the queer is emphasised.

## Can I wipe my hands now please?

In conclusion, queer readings can facilitate a reading of the most mainstream repre-sentations from the most marginalised of positions. Queer readings can expose the constructedness of things that appear transparent, set and given. As such by some guardians of the centre ground of 'meaning', especially in the arts, queer readings can be seen as threatening because they allow a section of the fringes into the mainstream. Soap coyly invites, and some indeed flash their ankles and welcome, queer hands, covered in the residue of queer as they are, to smear across the surface of their representations, discovering the manufacture of the image and enjoying the tactile experience.

By linking communication theory, bardic function, the ex-nominated and socio-centrality, I have attempted to create a description (an invention?) of a way of looking that queer uses in order to penetrate mainstream images. Queer readings can be a very powerful tool in a critical arsenal that seeks to enjoy mainstream texts, but also to look at their inherent paradoxes (I'm thinking here of the way that it is impossible to represent a queer in soap, yet it is quite evident that they exist within them), and just as important, to celebrate these multiplicities.

Queer readings can be helpful in disentangling the mesh of representation with a side-long glance and a playful attitude. Reading in this way is figured as an active process of manoeuvring around a representation exposing its innards, its joins and hidden recesses. If we go back to the metaphor of the bar of soap in the bath, even when the hand is rinsed and the residue of queer has been laundered away, queer still has its effects – there is a scent, a dampness, a clarity, and a smoothness to the skin. Queerness and soap – kind to your skin.

You may now wash your hands.

## Bibliography

Ang, I., *Watching Dallas, Soap Opera and the Melodramatic Imagination.* (London: Methuen) 1985.

Barthes, R., *Mythologies.* (London: Paladin) 1973.

Beemyn, B. & Eliason, M.(Eds.), *Queer Studies. A Lesbian, Gay, Bisexual, and Transgender Anthology.* (New York: NYP) 1996.

Buckingham, D., *Public Secrets, Eastenders & Its Audience.* (London: BFI) 1987.

Duberman, M.(Ed.), *Queer Representations, Reading Lives, Reading Culture.* (New York: NYP)1997.

Fiske, J. & Hartley, J., *Reading Television.* (London: Methuen) 1978.

Fiske, J., *Introduction to Communication Studies.* (London: Routledge) 1990.

Geraghty, C., *Women and Soap Opera: A Study of Prime Time Soaps.* (Cambridge: Polity Press) 1991.

Intintoli, M.J., *Taking Soaps Seriously, The World of Guiding Light.* (New York: Praeger) 1984.

Jagose, A., *Queer Theory an Introduction.* (New York: NYP) 1996.

McCarthy, V., 'Passion Play' in *Attitude* April 1999.

Mohr, R. ,*Gay Ideas, Outing and Other Controversies.* (Boston: Beacon) 1992.

Nochimson, M., *No End to Her. Soap Opera and the Female Subject.* (Berkeley: University of California Press) 1992.

O'Sullivan et al., *Key Concepts in Communication.* (London: Routledge) 1983.

Pattenden, S., 'Officer & Gentleman'. in *Attitude* May 1998.

Spargo, T., *Foucault and Queer Theory.* (Cambridge: Icon)

Tulloch, J., *Television Drama. Agency, Audience and Myth.* (London: Routledge) 1990.

Tunstall, J., *Television Producers.* (London: Routledge) 1993.

For gay magazine interviews and discussions of gay characters in soap:

Parkes, J.C., What's a Tiff between lovers? in *Gay Times* Nov. 1996

Patterson, S., And God created Adam. in *Attitude* May 1999

Radclyffe, M., A slice of this life. in *Gay Times* Aug. 1997.

Todd, M., Does seeing gays in soaps make coming out easier? in *Attitude* Sept. 1996

Todd, M., Soap Studs. in *Attitude* Nov. 1996

White, M., Mad about the boy in *Attitude* April 1998

Wilson, A., Streetwise in *Attitude* Nov. 1996.

# Framing and Reframing the 'Other'

## The Black Explorer: Female Identity in Black Feminist Drama on British Television in 1992

Claire Tylee

### Introduction

Breaking fresh ground by discussing Black feminist television drama, this chapter examines three screenplays from the co-called 'Columbus Year' of 1992, which were written by Meera Syal, Jackie Kay and Winsome Pinnock, dramatists already known for their theatre writing. As Black British feminists their art shares common political aims with the African-American authors who inspired them, yet it forms a distinct tradition from American womanism. Within a specifically British social context, these aims contest stereotypes of Blackwomanhood, give Blackwomen a voice, and display the effects of racism. Taking three television genres familiar to British audiences, the gothic romance, the political thriller and the true-crime history, these dramas subvert generic conventions and thus the cultural expectations and social values which these support. All three plays feature liminal Black female protagonists who move between cultures, exploring them on behalf of the (mixed) audience. Solidarity for women across boundaries seems to exclude men from this embrace, unlike dominant TV practice which places white male desire centre screen. The narrative foregrounds the Black female's quest for autonomy, with plot resolution expressed in terms of mutually supportive female relationships rather than heterosexual dominance and subordination.

This paper opens an exploration of the contribution made by Black British feminists to British TV – 1992 was a landmark year for the commissioning of work connected with Black experience because of the celebrations of the so-called Columbus anniversary (500 years since the 'discovery' of the New World led to the Black diaspora). As a result, 1992 was also a significant year in the TV broadcast of work by Black British women dramatists who had formerly written for the stage, with three major BBC2 transmissions by award-winning writers. These were Winsome Pinnock's script for the screenplay, *Bitter Harvest*, Meera Syal's melodrama, *My Sister-Wife*, and Jackie Kay's poetry-documentary,

*Twice Through the Heart*. However, that birth of Black feminist television drama was shortlived; television programmers failed to encourage its growth.

In the terms of Trevor Griffith's dictum that 'The plays which get deepest are plays which are aware of their own conventions, or other conventions, and which somehow or other manage to spring the unexpected within those conventions' [Griffiths 1982: 37], these plays are all 'deep plays'. They spring the unexpected. The three scripts all subvert dominant generic conventions which perpetuate the silence, invisibility or stereotyping of Blackwomen. The three generic forms employed are the gothic romance, the political thriller, and the true-crime history. By unexpectedly inserting an articulate, autonomous Blackwoman as the protagonist in generic forms which normally ignore or subordinate Blackwomen, the scripts drive a wedge into the social expectations validated by the formal conventions and open up questions of gender, culture and race.

By their strategy of undermining the audience's formulaic expectations, Pinnock, Kay and Syal question the wider social preconceptions and norms which these television genres usually support. In particular, their dramas relate personal, family matters to larger political structures of male power and cross-cultural incomprehension. However, this not only affects ideological constructions of Blackwomanhood. It also challenges the dominant norms of white male authority, agency and primacy, which contribute to the present racial and patriarchal hegemony.[1] Since television controllers themselves benefit from that current order, that may explain why they did not further these dramatic developments [Blanchard 1982,p123].

The three Black British women dramatists I am going to consider were all born in 1961 and graduated from University in 1982 – 3. Meera Syal was born in a village near Wolverhampton of South Asian parents and read English and Drama at Manchester University for four years. There she discovered not only Germaine Greer, Kate Millett and Simone de Beauvoir, but also Amrit Wilson. Jackie Kay was born in Edinburgh to a Scottish mother and African father, but was adopted and raised in Glasgow. She read English at Stirling University, where she studied Audre Lord as well as Marilyn French, Toni Morrison and Alice Walker. Winsome Pinnock was born in Islington to Jamaican parents. She read English and Drama at Goldsmith's before taking an MA at Birkbeck, University of London; her plays cite Simone de Beauvoir and Alice Walker. The three were known for their theatre-work before they wrote for radio or television, and they all benefitted from, and contributed to, writers' workshops and feminist support networks in the 1980s. Although they have never directly collaborated and have forged different career patterns, the three women have mutual connections and these networks link them together, relating fringe cultures to dominant and elite cultures in British society.[2]

Jackie Kay's two stage-plays, *Chiarascuro* (1986) and *Twice Over* (1988) were first given rehearsed readings by Gay Sweatshop, performed by the Theatre of Black Women and published by Methuen. Her poetry cycle, *The Adoption Papers*, was dramatised and broadcast in the BBC Radio 3 series, *Drama Now* in 1990, before being published by BloodAxe. It won the Eric Gregory Award 1991. *Twice Through the Heart*, a work commissioned for television that was broadcast and published by the BBC in 1992, has since been used as a libretto by Mark-Anthony Turnage and performed in an adapted version by the English National Opera Company at Snape, the Queen Elizabeth Hall and the

London Coliseum in the summer and autumn of 1997. Kay remains best-known for her poetry, but she has also published a successful novel, *Trumpet* (1998).

Winsome Pinnock worked briefly as an actress after graduating but also joined the Royal Court's Young Writers' Group, where she was tutored by Hanif Kureishi. A number of her plays received their first airing in the Royal Court's Theatre Upstairs. The Women's Playhouse Trust took up *A Hero's Welcome* (1986), the Women's Theatre Group toured with *Picture Palace* (1988), and she won the George Devine Award in 1991 for *Talking in Tongues*. Her best-known play remains *Leave Taking* which premiered at Liverpool Playhouse in 1987 and was toured by The Royal National Theatre during 1994 – 5. Her theatre work has been published by Nick Hern, Methuen and Faber. *A Hero's Welcome* was published by Aurora Metro in a volume alongside Meera Syal's television drama, *My Sister-Wife* (1993). Pinnock has been employed to provide episodes for *South of the Border* (BBC1 1988) and *Chalkface* (BBC2 1991), as well as to collaborate on the television screenplay *Bitter Harvest* with Charles Pattinson for BBC2 in 1992, but none of her television work has been published. She is at present a visiting fellow at Cambridge University.

Meera Syal first came to prominence in 1983, performing her one-woman stage-play co-written with Jacquie Shapiro – *One of Us*, about an Indian girl from Birmingham, which won a Yorkshire TV Award. Since then her acting career has progressed hand-in-hand with her writing. Well-known as a writer and comic for the television cabarets *The Real McCoy* (from 10 May 1991 BBC1) and *Goodness Gracious Me* (from 21 January 1998 BBC2), she has appeared in various television and radio series such as *The Bill, Legal Affairs* and *Absolutely Fabulous*, as well as in films like *Heavenly Creatures* (1994) and Hanif Kueishi's *Sammie and Rosie Get Laid* (8 March 1990) She made her name in fringe theatre acting with groups such as Monstrous Regiment, then went on to classical theatre at Bristol Old Vic and the National Theatre, as well as taking part in the huge success of Caryl Churchill's *Serious Money* (1987), which transferred from the Royal Court to the West End and off-Broadway. She has also written for television and radio series such as *Citizens, Black Silk* (7 November 1981) and *Tandoori Nights* (4 July 1985). However, her attempt at Black feminist popular television, a 3-part series about Southall Black Sisters called *Hungry Hearts* with an all-Asian cast, commissioned by Channel 4, was never finally funded. Undeterred she wrote the screen-play for the Black feminist cinema-film, *Bhaji on the Beach* in 1994.[3] Syal's only television work to be published is the play she wrote for BBC2 in 1992, *My Sister-Wife*, although she has published fiction, including the novel *Anita and Me* (1996) which won the Betty Trask Award, and had an accompanying audio-tape. A second novel, *Life Isn't All Ha Ha, Hee Hee*, was published in 1999.

This should make clear that these three women writers were already significant, experienced and successful writers for the stage, whose work has appeared in print and won awards, and who have moved from the fringe to mainstream, elite British culture. But their brief opportunity to transmit their own serious ideas on television in 1992 has not been repeated or developed. They have not been urged to build on that experience by writing serious drama for television again. Nor has any channel besides BBC2 taken them up or profited from this example. In 1994, BBC2 did broadcast an original play by the Black American woman writer, Bonnie Greer, *White Men Are Cracking Up*. And in

1998, BBC2's Screen 2 broadcast a play about a women's refuge by Tanika Gupta, *Flight*, at 10pm, with an all-black cast, including Meera Syal. So far as I have seen, though I may be wrong, this has been all; nothing on Channel 4, surprisingly, or ITV.[4]

Before analysing their plays I want to explain why I have grouped these three writers together as Black British; and what connections there are between 'Blackness' in Britain, and 'Blackness' elsewhere, such as the USA. What might Black British Feminism be said to constitute and how does it differ from, for instance, the womanism of African-American women writers such as Alice Walker or Toni Morrison? These are all issues connected with the significance of 1992 as a landmark year for Black awareness on British TV. One of my aims is to stress what has already become apparent from my introduction – the similarity but diversity between these three women, since this is a significant feature of Black British Feminism. This will help to explain why I think it is of interest to consider their writing together, despite the variety of forms (or genres) in which they wrote drama for BBC2 television in 1992.

So, what is meant by Black British feminism? It is clearly not up to me to define or to redefine 'Blackness' and recent women's studies conferences have evidenced a generational difference in the use of the concept and the recirculation of discourses of Blackness.[5] Like 'feminisms' this is a concept and practice continually in process. However, I am going to draw on self-definitions offered in the 1980s by various collectives such as Organisation of Women of African and Asian Descent (1978 – 83), the Centre for Contemporary Cultural Studies at Birmingham University, which published *The Empire Strikes Back* in 1982, and the editorial groups which produced *Many Voices One Chant* (a special number of *Feminist Review*) in 1984, *Heart of the Race* in 1985, and *Charting the Journey* in 1988. (Like her colleague, Pratibha Parmar, Jackie Kay was directly involved in several of these projects.[6])

These various groups defined Black British Women as women born in Britain of African or South Asian descent (that is, with a forebear (parent or grandparent) from the Caribbean, Africa or South Asia), who suffer similar exploitation/oppression/silencing because of the racism that results from the imperial colonisation which also accounts for their being in Britain. However, as *Charting the Journey* makes clear, there is no one structure or organisation that unites Black British women beyond the historical link – no unitary voice; no one cultural background, any more than there is for women in general other than the socially constructed and contested concept of Woman (Knowles & Mercer p64; Riley). But they share a similar social position and political beliefs [Grewal 1988: 1 – 6]. So the situation of Black Britons is different from that of African Americans, who have a common racial background, the historical experience of slavery and legally enforced segregation, and 2 – 300 years of being rooted in the USA. In particular, the spiritual beliefs and practices which helped African American women survive that history and which inform their distinctive womanism are not apparent in British writing. Yet Black British women's experience of racial discrimination and of the double subordination of race and gender is similar to the experience of African American women. Feminists in both groups see a need for specific political action and the American example opened opportunities for the British. As writers, particularly for the theatre, the three Black British women whom I am considering here share with their better-known American sisters a feminist concern with the silencing and invisibility of women, especially Blackwomen, and especially lesbians.

The year 1992 was a landmark year because of the 500th Columbus Anniversary of 1492 which had led to the diaspora of African and Asian people. The massive publicity in the USA and Spanish/Portuguese speaking countries spilt over into UK. It encouraged publishers to finance trips to UK by a number of African-American women writers, such as Toni Morrison, Alice Walker, Maya Angelou and June Jordan, to speak at events in London associated with the C. L. R. James festival and the production of Ntozake Shange's *The Space Love Demands* starring the performance poet, Jean Binta Breeze. Blackwomen in Britain have spoken of the influence of such writers on their own work, and the encouragement that they gained from their example, despite the fact that only one of these was a dramatist. Thus for instance, Winsome Pinnock has claimed that the hardest thing of all is creating a tradition of Black British women's playwriting, 'There is no Black Caryl Churchill.' In that absence, she says her most significant influence is Toni Morrison, 'who writes with such compassion and attack' [McFerran 1991]. Jackie Kay has spoken of the influence of Shange because she showed how poetry could work as theatre [Kay 1987: 124 – 5]. The main effect of the African American women-writers came from their shared enterprise of placing Blackwomen and their experiences at centre-stage. But it is important that the interest in American Blackwomen should not have the result of displacing British Blackwomen's history, further reinforcing their silence. As Barbara Burford said, 'Many of us obviously share a common ground with the Black American experience, but many do not . . .we have our own voices, and our own concerns'; she then demanded that 'all these voices, in their rainbow array, be heard' [Burford 1988: 98].

The aim of the Black British playwrights to give women a voice, was held in common with white feminists of the 1970s and 1980s [WARSAG 1982]. However, Blackwomen are muted twice over, made invisible twice over, both because of gender and race. Pinnock has spoken of the importance of giving voice to an immigrant Jamaican woman 'who is largely invisible in our society because of her colour and her class' [Croall 1994, p3]; Syal has written of the need to give a voice to Asian women in Britain who are silenced by a society that regards them as second-class [Syal 1990a]. Kay has been associated with editorial collectives who have challenged white academic feminists for failing to recognise the significance attached to colour and race, as well as to class; one of her own poems asserts: 'We are not all sisters under the same moon, [Parmar 1985: p58-9]. Above all, these playwrights share a desire to write beyond the stereotypes of Blackwomanhood found in the work both of white writers (of either sex) and of Black male writers – stereotypes of, for instance, passive Asian women oppressed within their family, or of strong, dominant Afro-Caribbean women, heads of households. One of their main aims has been to redefine the very idea of 'Blackness' and 'Black womanhood' without imposing a new, spurious unity.

The other problem they all tackle in various ways is one shared by Black Britons of both sexes, the sense of being between two cultures and not being sure where they belong, 'where home is', as Winsome Pinnock puts it [Croall 1994: p3][7]. Syal's first play concerned the sense of self-dissolution that comes with being pulled by both cultures, and of the need to reconcile them and find strength in difference. For someone of mixed race like Jackie Kay this was a particularly painful problem which involved recognising rejection in order to find pride in her identity, especially as a lesbian. In fact, their intimate knowledge of two cultures as well as their sense of standing outside them, makes these three writers especially suited to translating between cultures in their work. This hybrid-power seems to me of prime importance in the radically pluralist society that Britain needs to become [Donald & Rattansi 1992: 5]. As she herself recognises, Syal is a role-model both as actress and writer: 'People of my age feel that having two cultures is a positive advantage. I love being a British Asian . . . I'm an outsider so I can see both my culture and my host culture more clearly than somebody who grew up in either one' [Syal 1991: 120; Syal 1990b].

The works that Syal, Kay and Pinnock created for BBC2 in 1992 all involve Black liminal protagonists with whom a mass audience can identify, who cross to explore an unknown, hidden, muted culture and in the process question the dominant culture from which they set out. So, although the works satisfy a mass audience's demand for pleasure, the pleasure from suspense, the thrill of violence, the titillation of vicarious sex, and from gratified curiosity about the unknown, they can also be seen to conform to Kay's views about writing as a political activity with 'a tremendous power to effect change' not through polemics but through the transforming potential of the poet's vision of 'a world that wouldn't be racist or sexist or homophobic,' [Wilson 1990: 124]. In her view, for people to acknowledge and embrace differences between them they have to undergo a journey, a journey that involves recognising and understanding and shock and anger [125]. In their BBC2 dramas, these writers seduce the audience to go on such a journey, accompanying the liminal protagonist from a known, safe world of preconceptions to a place where these preconceptions are upset. Obviously, using the metaphor of a journey in the context of 1992 seems particularly appropriate, but for the audience the journey is also a mirror version of itself – through the looking-glass to another culture, to see oneself from the other side. The particular creative power of these dramatists' vision is to construct the television screen as a looking-glass for both Black and white audiences, where we all can see ourselves and others from a different point of view.[8] This is one reason why it is important that Black writing not be ghettoised in 'ethnic programming'.

Having so far stressed the common project shared by these dramatists, I want now to indicate the differences between their works and the ways in which they set about achieving their common end. The most obvious differences are due to the different generic forms they chose to work in. These forms, belonging to the dominant culture, are constraining for marginalised perspectives but, precisely because they operate with recognisable conventions, they offer opportunities to challenge those conventions and the prevalent values they normally transmit [*cf* Dubrow 1982: 3 – 4]. I'm going to deal with the three works in the order in which they were broadcast, starting with Meera Syal's play, *My Sister-Wife*, moving on to Winsome Pinnock's screenplay, *Bitter Harvest*, and finally discussing the work which I found the most subversive, Jackie Kay's *Twice Through the Heart*, billed as a poetry documentary.

*My Sister-Wife*, which starred Meera Syal herself in the leading role as the liminal protagonist, at first conforms to the formulae of the gothic romance such as *Rebecca* or

*Jane Eyre*, where the second wife is haunted by the [idea of the] first wife. The main difference is that this romance is set in a moslem context in Britain, and the big house essential to such a plot has became a harem in which the husband actually has two wives concurrently – the second wife, a Europeanised Asian business-woman called Farah, has to accommodate herself to sharing with the first wife, Maryam. Although she speaks English fluently, Maryam has come straight from Pakistan and still speaks Urdu with Asif (the husband), with her mother-in-law (Sabia), and with Maryam's two young daughters, who also all live in the house. Farah doesn't understand Urdu; Sabia will not speak English. Syal makes use of the opportunities offered by television to present some of the Urdu to an English-speaking audience by means of subtitles, subversively demonstrating Sabia's powers of ridicule in both English and Urdu.

Gradually the audience's sympathies become divided. With each other as example, Maryam becomes more vocal and self-assertive, Farah withdraws into silence and madness. Thus Syal follows recent subversions of the traditional gothic genre by feminist writers such as Jean Rhys in *Wide Sargasso Sea* and Fay Weldon in *Life and Loves of a She-Devil*, in giving voice to the previously silenced first wife. She also follows their example by making the heroine the gothic, murderous monster, Bronte's 'madwoman in the attic' to whom Syal's play explicitly alludes. However, as a Black feminist, she goes one step further. Not only is the madwoman the second rather than the first wife, but it is the husband, Asif, who is murdered, and the two women end bound together as sisters. Thus Syal has thoroughly disrupted not only the traditional gothic romance conventions, (of the evil first wife) but also the modern white feminist reworkings (in which the monstrous first wife is made the protagonist). Here the gothic self and 'dark' Other are finally united, despite their differences, across a cultural divide and in the face of a sympathetically portrayed, but then eradicated, patriarchy.

Syal was exceptional in having near total control over her play – not only writing it and starring in it, but also choosing a female producer, Ruth Baumgarten, who supported her feminist aims. Winsome Pinnock's experience was quite different.[9] Pinnock was brought in at a late stage on a project initiated by the television producer, Charles Pattinson, as part of the anti-Columbus response to the 1992 fanfare about the Year of Discovery. Prompted by a report in *Americas Watch* in 1989, Pattinson had travelled to the Dominican Republic (in the Caribbean) on a brief research trip as part of a scheme to expose the plight of migrant Haitian workers there. He then sketched out a story based on attempts by aid-workers to improve conditions for present-day sugar-slaves in the canefields [Smith 1992]. This first draft was sent to Josette Simon (who at the time had just received the Olivier Best Actress Award for her role in Arthur Miller's *After the Fall* on the stage at the National Theatre). After Simon had agreed to take the lead-role, Pinnock was called in to co-author the screenplay. The overall plot resembles the film-script of a feature film Simon had already starred in with Peter Postlethwaite, *A Child From the South* (1991). Ironically Pattinson's film-team encountered the same kind of problems working in a police-state that the makers of *A Child From the South* had encountered in South Africa. In order to get permission to film, Pattinson had to convince the authorities that he was making a political thriller. This to some extent explains the unsatisfactory end-result, since the film attempts to reveal the sordid conditions behind the glossy facade presented to tourists, when political thrillers of the James Bond variety actually trade on that gloss and the film itself has the cheap but glossy finish of a tourism commercial or travel show.

Furthermore, having Pinnock in as co-writer, whilst it produced the credible protagonist portrayed by Simon, complicated the political aims of the film. According to Pinnock, there was real collaboration between herself and Pattinson – they talked a great deal and divided the scenes up between them.[10] However, Pinnock did not get to go to the Dominican Republic. The resultant film has two main stories. One concerns the relation between Vivienne (Josette Simon), a recent law-graduate, and her parents, a white-English mother and Black Trinidadian father, a lawyer who had abandoned the family and country of his birth. It is against his wishes that Vivienne goes to the Dominican Republic as an aid-worker to discover her roots in the Caribbean.[11] The film begins when the parents fly out there from London after she has gone missing, and the search provokes confrontation between the parents. This plot resembles Pinnock's exposition both of generational conflict in *Leave Taking* and of the destructive anger underlying sexual relations between Blacks and Whites in her play, *Talking in Tongues*, of which she said: 'I want to show how racism can destroy relationships' [McFerran 1991].

Clearly Pinnock's contribution – this plot concerning the antagonisms in a mixed-race family – sits uncomfortably alongside the political-thriller aspect of the film, where Vivienne gets caught up in a workers' strike which results in the murder of one of the Haitian strikers. Whilst Vivienne's story reveals the exploitation of the Haitian cane-workers, her father's story reveals the sordid side of the sex-tourist industry. Neither she nor her father sufficiently resembles the usual hero of a thriller to successfully play on thriller conventions, and the plots end by humiliating both of them, fetishising the black female body along the way. The film was certainly a brave enterprise, and Pinnock was more than a hack-writer. But I suspect that its unsatisfactoriness stems from the differing political aims of Pattinson and Pinnock. I don't believe this would have been solved merely had the collaboration started earlier in the creative process; the differences are too fundamental.

By contrast, I found Jackie Kay's *Twice Through the Heart* (19 June 1992 BBC2) more successful, and the most subversive of the three works. It was commissioned by Peter Symes as part of a series of six half-hour films, *Words on Film*, which aimed to combine verse and documentary in the tradition of Auden's *Night Mail* (1936). This is a form which has been recently developed with great creativity by Tony Harrison. The idea was for the poet to choose a subject and then to research it together with a film-director, but the film was to be driven by the poet's imagination. The poet and director worked together on the filming and editing, juxtaposing lines of poetry with filmic images [Symes 1996: 9 -12]

Kay was the only female poet of the six, and she chose a topic of current feminist controversy – the jailing of women for killing their abusive partners in self-defence. In fact this was an issue of particular concern to Black feminist groups like the Southall Sisters, who became nationally associated with the campaign to free women in that position, especially Asian wives. But Kay took the case of a white British woman, Ethel

Amelia Rossiter, aged 63, jailed in February 1988 for the murder of her husband the previous July by stabbing him 'twice through the heart'. Although Rossiter was the main person involved, she did not actually appear in the film. Instead, the protagonist was Kay herself, who appeared as a kind of parody of Edgar Lustgarten in a straight-faced satire of his television real-life crime' series. In those programmes, the 'true' tale of the known facts about a sensational historical crime, the criminal investigation and the subsequent trial, was ghoulishly introduced and narrated by Lustgarten, and dramatically re-enacted in a costume documentary.

In Kay's version, it was not so much the crime which was under investigation as the criminal justice system itself. Indeed her documentary challenges the very idea that the killing was a crime of murder. She used her poetry to imagine a subjectivity for Rossiter and explore how she had been silenced. Kay acted as a kind of Black-Scottish anthropologist journeying to the Heart of Darkness which lay in the Middle Temple of the English Legal Institution. Visually, the film ridiculed the court-ritual by a section on a legal-wig-making factory – which for feminists was a neat allusion to the photos and arguments in Virginia Woolf's anti-patriarchal polemic, *Three Guineas* [Woolf 1992: 240 – 4]. Together with representing the jury by clothes on hangers, this provided a visual mockery of costume drama, where the pleasures of nostalgia and vicarious consumption tend to outweigh any serious interest in 'the facts'. Audibly the film contrasted the regional accents of Kay and Mrs Rossiter's daughter and neighbours with the upper-class voice of the judge. At the heart of the documentary was a non-verse interview with the barrister Helena Kennedy, who pointed out that the law was defined by men, and that Rossiter had suffered from their male way of approaching crime in terms of physical equals.[12] Kay used her outsider status as not only a Scot, and a woman, but also as racially Other, to present herself in the topsy-turvy role of a scientific investigator, rather like David Attenborough, and to position her audience as co-explorers of the strange events, customs and practices of two arcane institutions – patriarchal marriage and the Law.

How did the dramatists try to achieve the 'mixed audience' Pinnock desired? [Stephenson 1997: 53]. How did they set up their central 'Black female explorer' as a figure majority white audiences could identify with sufficiently to accompany on their journey to shock and anger, without alienating Black audiences? [Whitemore 1982, 103]. And how, at the end of the journey, did the scripts encourage mutual acknowledgement and what Jackie Kay called 'the embrace and acceptance of differences'? [Wilson 1990: 124]

Beginning before the titles all three programmes entice any viewer in with a succession of brief, enigmatic shots and mysterious sound; these tantalisingly crystalise the heroine's mental state at a determining moment of crisis in her life. In *My Sister-Wife* it is the image of her husband's infidelity, which provokes Farah's madness and act of murder; in *Bitter Harvest*, it is Viv's intervention in the confrontation between the strikers and their armed guards, which precipitates a killing and her breakdown; in *Twice Through the Heart* it is the final physical abuse, that traps Rossiter into her desperate act of self-defensive stabbing. The stories after the titles start elsewhere. *My Sister-Wife* begins with Farah sitting by her bed watching an extravagant Indian love-film on television with her best friend, Poppy, a journalist (Harriet Bagnall). Poppy speaks first and her wit and kindness provide an ironic point of identification for (white) viewers as she sympathetically but uncomprehendingly tries to support Farah in her transition to 'traditional' moslem ways. Farah has a Midlands accent and wears European clothes; played by Syal she would immediately feel familiar to television audiences. As Pinnock put it simply (with regard to *Leave Taking*):

> When you watch a performance of a play you are forced to identify with whoever is the heroine at that point. So people were identifying with someone who was seemingly 'other' . . . irrespective of their class or race. But that's what theatre does . . . people can identify with it, even if it's alien to their culture. You can enter different worlds [Stephenson 1997: 48-9]

The inter-cultural scene has been set – Indian, white English and Black British, offering three points of recognition when Poppy meets the silent Maryam (Shaheen Khan) at Farah's wedding. There are three journeys in the course of the play, although Poppy's is of less importance; by the end of the play Farah has moved from close friendship with Poppy to the sisterly arms of Maryam, in another bedroom.

The narrative of *Bitter Harvest* begins as a plane's wheels touch down and a couple arrive at a foreign airport, a holiday experience instantly recognisable. A sub-title 'Dominican Republic' appears on the screen, as Miriam and John Johnson reach the barrier. Played by Sue Johnston and Rudolph Walker, their faces would also be instantly recognised by television audiences. Woddis describes Johnston as 'a hugely popular actress who, as Sheila Grant in *Brookside*, Channel 4's Liverpool soap, attained something akin to folk heroine status', her performances 'radiating a remarkable emotional truth' [Woddis 1991, 158]. Her presence ensured audience empathy as again a three-way inter-racial situation was established – white English, Trinidadian British and their Black British daughter, in an alien environment where none of them feels at home. Although all three take the audience on a separate journey, John's destination is unclear and Miriam has already established hers. It is Vivienne's which is fundamental as she learns that 'These are not your people' and what it means to share English values. The film ends as she finally faces up to her parents, clasped in her mother's arms and looking at her father over her mother's shoulder.

*Twice Through the Heart* is more compressed and, although filmed on location, not 'naturalist'. The 'heroine', Rossiter, is never seen and so the audience is denied any visual signifier of her race. The first words we hear are 'No Way Out', the thoughts Kay attributes to her, spoken by an offscreen actress. The next voice is the cheerful Glaswegian of Kay; before we see her face it is her comfortable shoes that provide the audience with human contact when confronted by the forbidding aspect of the Old Bailey with its stone figure of Justice. It is Kay who guides the viewer on a journey through the rigmarole and ritual of court and prison, diminishing their status and permitting Rossiter's daughter to address us in her own words. The piece ends with Rossiter, aged 67, still inside, and Pat outside Bullwood Hall Prison with her hands at the grill: 'I just feel, well, you know – it's – ' and Pat turns her face away as her voice chokes. Clearly the whole system is preventing the hugs we saw at the end of *My Sister Wife* and *Bitter Harvest*, and which a female audience vicariously identifies with – the loving care for an Other 'on the wrong side of the law', which all three scripts them-selves demonstrate.

In none of the dramas is any attempt made to embrace the patriarchal white male, the true polar opposite of the Blackwoman and source of difference. He remains the exclud-ed Other. Such hugs as there are seem to be to construct solidarity between women, both characters and audience, in defiance of him and his values, whether it be sexual tourism, corrupt foreign aid, the vilification of women who kill, or the arrogance of men-in-the-media satirised at the party in Syal's *My Sister-Wife*. That catalogue is not exhaustive, which suggests there remains plenty of scope for Black feminist drama on television. Opportunities to develop a body of Black feminist drama might widen its range to include differences amongst white males. Whether any of these differences could be not only acknowledged but also embraced remains to be seen.

## Notes

1   In the following discussion I am adapting the theoretical analysis of feminist fiction provided by Anne Cranny-Francis in which, referring to Bakhtin, she argues that feminist modifications to generic fiction 'renegotiate the contemporary social/ideological formation – bourgeois, patriarchal, white supremacist society – by constructing for readers a reading position which is discursively at odds with that construct-ed by texts which simply [re]construct the dominant form' (Cranny-Francis 1990: 195-6). Feminist writers for television offer to a mass audience subversive audience-positions.

2   To give an example of such networks: as co-founder of the Theatre of Black Women, Bernardine Evaristo contributed introductions to the collection of plays edited by Kadija George which contained plays by Syal and Pinnock (George 1993). In 1986 Evaristo had worked with Kay on the first draft of her play *Chiarascuro*. Evaristo and Syal both contributed to Maud Sulter's anthology *Passion* (1990) and Sulter interviewed Alice Walker for the collection co-edited by Kay, *Charting the Journey* (1988).

3   Funded by Channel 4, this was the first feature film to be directed in UK by a British Asian woman, Gurinder Chadha.. She had cut her teeth on a 10 minute short for Channel 4, also written by Meera Syal: *A Nice Arrangement*, shown at Cannes in 1991.

4.  Since 1991, Gupta has contributed episodes to *Grange Hill* and *Kiss Me Quick*, but otherwise what Meera Syal said is still true: 'There aren't any Indian women writing for television, and the few parts that do exist are written by people who don't know anything about us' [Syal 1991, p115].

5   For instance at the plenary session of the Women's History Network National Conference at Manchester in 1996.

6   Pratibha Parmar, the British Indian film-maker who worked on *Warrior Marks* with Alice Walker, was a key figure in these collaborations, contributing to *The Empire Strikes Back* and co-editing *Many Voices One Chant*, *Charting the Journey* and other Blackwomen's publications such as *A Dangerous Knowing* (1985). Important theoretical essays from these works have been reprinted in Heidi Safia Mirza's *Black British Feminism: a reader* and I draw on these in the following paragraph.

7   For instance, this anxiety is explored in Rosalie John Baptiste's play, *No Place Like Home*, where Marcie complains: 'But I'm not English, Mam, and I'm not Dominican either' [Baptiste 1987:144]; see also Lauretta Ngcobo on 'unbelonging' [Ngcobo 1987:9-10] and Naz Rassool about the importance of a knowledge of the past [Mirza1997: 190].

8  British theatre audiences are notoriously white and middle-class, whereas writers such as Pinnock want their work to be seen by 'a mix of people' [Stephenson 1997: 53] – television obviously makes that more possible, and it is clear that Kay, Syal and Pinnock opened their work to such a mix in 1992.

9  It demonstrates the importance of Black women being not only writers but producers and editors too, if television is not to be entirely 'organised around male erotic privilege' [Clayton 1982, p 153 – quoting from Laura Mulvey].

10  Answer to question at Platform Event, Pinnock interviewed by Jack Bentley, National Theatre 13 January 1995.

11  Part of the film's dramatic license involves skirting over why, to discover her British West Indian roots in Trinidad, an intelligent woman should go to an ex-Spanish colony 1000 miles away (like an Irish -American searching for his roots in France).

12 Kennedy was at the time preparing her book on this subject:

## Bibliography

Amos, Valerie, Gail Lewis, Amina Mama and Pratibha Parmar edd, *Many Voices One Chant, Feminist Review 17*, London, July 1984.

Baptiste, Roselie John, *No Place Like Home* in Considine, Ann and Robyn Slovo edd., *Dead Proud from Second Wave Young Women Playwrights* London, Women's Press, 1987, pp141-156.

Blanchard, Simon and David Morley edd., *What's Chanel Four? An Alternative Report*, London, Comedia, 1982.

Bryan, Beverley, Stella Dadzie and Suzanne Scarfe, edd., *The Heart of the Race: Black Women's Lives in Britain*, London, Virago 1985.

Burford, Barbara ' "...and a star to steer her by"?' in Grewal, Shabnam, Jackie Kay, Liliane Landor, Gail Lewis and Pratibha Parmar edd., *Charting the Journey: Writings by Black and Third World Women*, London: Sheba, 1988, pp97-9.

Centre for Contemporary Cultural Studies edd., *The Empire Strikes Back: Race and Racism in 70's Britain* London, Hutchinson, 1982.

Clayton, Sue, 'Cherchez la Femme' in Blanchard, Simon and David Morley edd., *What's Chanel Four? An Alternative Report*, London, Comedia, 1982, pp 151-6.

Cranny-Francis, Anne, *Feminist Fiction: Feminist Uses of Generic Fiction*, London, Polity, 1990.

Croall, John, 'Acknowledging your own history', *Stagewrite*, Winter 1994, p3.

Donald, James and Ali Rattansi eds., *'Race', Culture and Difference*, London, Sage, 1992.

Dubrow, Heather, *Genre*, London, Methuen, 1982.

Du Maurier, Daphne, *Rebecca*, London, Gollancz, 1938.

George, Kadija ed., *Six Plays by Black and Asian Women Writers*, London, Aurora Metro, 1993.

Grewal, Shabnam, Jackie Kay, Liliane Landor, Gail Lewis and Pratibha Parmar edd., *Charting the Journey: Writings by Black and Third World Women* London: Sheba, 1988.

Griffiths, Trevor 'Countering Consent' in Pike, Frank ed., *Ah! Mischief; The Writer and Television*, London, Faber, 1982 pp30-40.

Harwood, Kate ed., *New Run: New Plays by New Writers*, London, Hearn, 1989.

Kay, Jackie, *Chiarascuro* in Jill Davis ed., *Lesbian Plays*, London, Metheun 1987.

_____ *Adoption Papers*, Newcastle, Bloodaxe, 1991.

_____ *Twice Through the Heart* in Symes, Peter ed., *Words on Film*, London, BBC Education, 1992.

Kennedy, Helena, *Eve Was Framed: Women and British Justice*, London, Chatto, 1993.

Knowles, Caroline and Sharmila Mercer, 'Feminism and Anti-racism' in James Donald & Ali Rattansi edd., *'Race', Culture and Difference*, London, Sage, 1992, pp104-125.

McFerran, Ann, 'Black Women at Centre Stage', *Independent on Sunday*, 23 June 1991, p19.

Mirza, Heidi Safia ed., *Black British Feminism: a reader*, London, Routledge, 1997.

Ngcobo, Lauretta, *Let It Be Told: Black Women Writers in Britain*, London, Pluto, 1987.

Parmar, Pratibha and Sona Osman edd., *A Dangerous Knowing: Four Black Woman Poets*, London, Sheba, 1985.

Pike, Frank ed., *Ah! Mischief; The Writer and Television*, London, Faber, 1982.

Pinnock, Winsome, *Leave Taking*, in Harwood, Kate ed., *New Run: New Plays by New Writers*, London, Hearn, 1989, pp139-189.

Rhys, Jean, *Wide Sargasso Sea*, Harmondsworth, Penguin, 1968.

Riley, Denise, *Am I That Woman?*, London, Macmillan, 1983.

Shange, Ntozake, *The Space Love Demands* in *Shange: Plays 1*, London, Methuen Drama, 1992.

Smith, Rupert, 'Strangers in "Paradise"' in *Radio Times*, London, 18-24 July 1992, pp22-4.

Stephenson, Heidi and Natasha Langridge edd., *Rage and Reason: Women Playwrights on Playwriting*, London, Methuen Drama, 1997.

Syal, Meera, 'Finding My Voice' (1990a) in Maud Sulter ed., *Passion: Discourses on Blackwomen's Creativity*, Hebden Bridge, Urban Fox, 1990, pp57-61.

_____ in interview (1990b), *Plays and Players*, London, March 1990, pp14-16.

_____ 'I feel I've Got a Niche Here' in Woddis, Carole ed., *'Sheer Bloody Magic'; Conversations with Actresses*, London, Virago, 1991, pp111-124.

_____ *My Sister-Wife* in George, Kadija ed., *Six Plays by Black and Asian Women Writers*, London, Aurora Metro, 1993.

Symes, Peter ed., *Words on Film*, London, BBC Education, 1992.

_____ 'Poetry and Documentary Film' in *Crosscuts Programme*, London, Royal Festival Hall, 3 November 1996, pp9-12.

Weldon, Fay *Life and Loves of a She-Devil*, London, Hodder, 1983.

Whitemore, Hugh, 'Word into Image: Television Dramatisation' in Pike, Frank ed., *Ah! Mischief; The Writer and Television*, London, Faber, 1982.

Wilson, Amrit, *Finding a Voice*, London, Virago, 1978.

Wilson, Rebecca and Gillean Somerville-Arjat, *Sleeping With Monsters: Conversations with Scottish and Irish Woman Poets*, Edinburgh, Polygon, 1990.

Woddis, Carole ed.,*'Sheer Bloody Magic': Conversations with Actresses*, London, Virago, 1991.

Women's Advisory and Referral Service Action Group, 'Opportunity Knocks (but not very hard)' in Blanchard, Simon and David Morley edd., *What's Channel Four? An Alternative Report*, London, Comedia, 1982, pp104-9.

Woolf, Virginia, *Three Guineas* (1938) in Morag Shiach ed., *A Room of One's Own & Three Guineas*, Oxford, Oxford University Press, 1992.

# Cultural Hybridity, Masculinity and Nostalgia in the TV Adaptation of *The Buddha of Suburbia*

Bruce Carson

## Preamble

This chapter considers how the television adaptation of *The Buddha of Suburbia* has mediated the issues of cultural identity that lie at the centre of Hanif Kureishi's first novel. The analysis uses a historical approach to explore how the novel's generic hybridity and its setting in the recent past of the 1970s make it an ideal vehicle for adaptation as a quality TV drama. The chapter argues that the majority of critical readings have privileged the novel and overlooked the importance of the television drama serial in shaping the audience's cultural memory of the text. The article also uses a mixture of feminist and psychoanalytic analysis to then explore how the TV adaptation's emphasis on a nostalgic *mise en scène* and the pleasures of spectatorship have intensified the novel's view of its adolescent hero, Karim Amir, as an Anglo-Asian exotic. However, this racial and sexual stereotyping of Karim does not always fully support the more Utopian notions of a post-colonial critic like Homi K. Bhabha. In particular, Bhabha's views that the historical processess of cultural hybridity help to break down essentialist conceptions of racial and ethnic identity. Finally, the analysis argues that the adaptation's explorations of identity and nostalgia can only be understood, like the novel, in relation to the sense of cultural crisis engendered by the anglocentric and market-driven policies of the Thatcher era.

The importance of Hanif Kureishi's status within the academy can be measured by how much his Anglo-Asian perspective has become a focal point for the application of post-colonial discourses[1] on issues of identity. He is perceived as offering strong evidence of contemporary Britain as containing a shifting diversity of cultures in which social identities are becoming more heterogeneous. His questioning of monolithic views of both Asian and British identity has led many critics to see his work as quintessentially hybrid. He has been spoken of as 'probably the most hybrid of the South Asian diaspora writers in Britain' (Jussawalla, 1997:17), and as a 'herald of hybridity' (Schoene, 1998: 117) where 'the hybridity of his insider/outsider point of view is unique' (Kaleta, 1998: 7).

The term hybridity is a concept that has become particularly associated with Homi K. Bhabha who is part of a post-colonial paradigm critiquing essentialist conceptions of racial and ethnic identity[2]. For Bhabha, this process is not so much a fusion of different cultural identities, as a Third Space in which people constantly explore their identities across the social boundaries of class, gender, sexuality, race, ethnicity, religion and nationality (Bhabha, 1994: 36-9). Kureishi's first novel *The Buddha of Suburbia* is seen as providing empirical evidence for Bhabha's theories of cultural identity[3]. Whilst most of the major characters in the novel show a desire to change their identities, it is Karim Amir, a suburban seventeen year old, who is at the centre of the narrative. A product of an English mother and an Indian father, his adolescent desires prompt him to leave the boredom of his lower middle class family life to explore the sexual and social boundaries of the 1970s. His journey of initiation takes him from the south London suburbs of his parental home to a successful acting career in the artistic and bohemian centres of London and New York. In his ability to cross social boundaries Karim is read by Berthold Schoene as 'a radically deconstructive presence', a hybrid character who 'inhabits an ethnicity-free-no-man's land' and as someone who fits perfectly into the 'in-between space' (Schoene, 1998:117) of Bhabha's Third Space.

The novel also uses a generic hybridity that links Kureishi's concern with issues of changing cultural identity to a wide range of comic and nostalgic pleasures. A fascination with the past has been a recurrent theme in British 'heritage' film and television, and the novel therefore formed an ideal vehicle for adaptation with its combination of social critique and ironic commentary on the shifting identities and lifestyles of the 1970s. This combination of elements made it a strong candidate for adaptation after its publication in 1990. After an initial reluctance, Hanif Kureishi agreed to adapt his work for television and collaborated with film director Roger Michell in writing the script (Kaleta, 1997: 113-114). The story was serialised as a four-part drama that was first shown on BBC 1 in 1993 (later repeated on BBC2 in 1995). Whilst a great deal has been written about the novel's articulation of issues of hybridity and identity, less exploration has been made of it as a literary historical adaptation for television. This is unsurprising given the critical and cultural prestige that still accrues to English literature. Television is perceived in taste terms as a 'low' cultural form, although drama is seen as one of its most prestigious areas, its creative status being assured by the cultural importance given to the writer.

## Television Drama, Race and Identity

Hanif Kureishi's success as a writer has helped to create him as a contemporary *auteur* with the 'postmodern' ability to range across a variety of 'high' and 'popular' cultural forms. However, his exploration of identity issues has not developed in isolation from the wider social, political and economic factors that have influenced film and television in the last twenty years. By the early 1980s the high cost of producing one-off TV dramas had led to the decline of one of the few areas that provided openings for new writers to challenge dominant representations of cultural and national identity. The search for alternative forms saw the new minority channel, Channel 4, pioneer a commercial strategy for high quality drama that involved the commissioning and financing of modestly budgeted feature films from independent film-makers. These films were given an initial cinematic release followed, at some later date, by a television transmission under the *Film on 4* banner. The approach was such a commercial success that it helped to push the new channel into the forefront of film-making in the UK and to provide career opportunities to promising young film-makers including those from racial and ethnic minority backgrounds[4]. Amongst the early successes was the work of Hanif Kureishi with his Oscar nominated filmscript for *My Beautiful Laundrette* (directed by Stephen Frears, 1985). Follow-up scripts for Frears' *Sammy and Rosie Get Laid* (1987), and for his own less successful directorial debut in *London Kills Me* (1989), helped to establish him as a rising talent. Kureishi's early success as a filmwriter made him an author in a collaborative medium that has traditionally seen its *auteur's* coming from the ranks of film directors. However, the blurring of aesthetic and economic boundaries between film and television in the 1980s were to Kureishi's benefit as a developing writer. It was a cultural status that was to lay the foundations for the later television adaptation of *The Buddha of Suburbia*.

Despite Hanif Kureishi's early success in film-making, there were still only a limited number of British Asian film-makers by the early 1990s, the most notable example being the director Gurinder Chadha with *Bhaji on the Beach* (1993), the first feature length film made by a British Asian woman. This was co-written with Meera Syal who had also written and performed in the television drama *My Sister-Wife* shown on BBC 2's *Screen 2* slot in 1992 (Ross, 1996: 46-50). These films were seen as part of a second 'moment' in black cultural politics that moved towards what Stuart Hall has called 'the extraordinary diversity of subjective positions, social experiences and cultural identities which compose the category black'(Hall, 1988: 28). It saw African Caribbean and British Asian filmmakers, the second generation children of migrants, use a variety of formal approaches to contest the dominant images of blacks and Asians in Britain. This was a politics of identity that challenged the very idea of racial, ethnic or national identity as fixed or stable whether it was black, Asian or British. The under- and mis-representation of diaspora communities still represents a sensitive issue especially in the context of continuing concerns about the widespread culture of racism in 1990s Britain. It is little surprise that in such an environment the few films or television programmes that are made by representatives of ethnic minority groups are seen as bearing a moral reponsibility not to render marginalised immigrant communities vulnerable to the judgemental gaze of white audiences. As an insider-outsider in both the Asian and white/dominant cultures, Hanif Kureishi's narrative strategies of not offering straightforwardly positive images of British Asians have often brought strong reactions from some members of the Asian community to his films. What Malik has called the 'pleasures of hybridity' (Malik, 1996: 202) saw objections to *My Beautiful Laundrette* as having a stereotypical view of Asians as homosexuals or corrupt businessmen (Ross, 1996: 111). Similarly strong reactions to *The Buddha of Suburbia's* language, drug taking and sex scenes were anticipated by the production team, and according to Kaleta some changes were made under pressure from the BBC. Despite these changes there were still tabloid criticisms and a refusal to show the adaptation in the USA (see Kaleta, 1998: 117-118).

## Kureishi, Thatcherism and the 1980s

This ability to court controversy can only be fully understood by also relating Kureishi's perspective to the conservative hegemony of the period. Kureishi's early films are both 'reactions to and critiques of the dynamics of Thatcherism' (Torrey Barber, 1993: 222) that came to dominate British society in the 1980s. One aspect of this ideological project was the fostering of a traditionally anglocentric vision of British nationalism that led many individuals from racial and ethnic minority backgrounds to feel excluded by this one nation philosophy. In contrast to this ethnic absolutism, Kureishi's two films with the director Stephen Frears showed a perspective on 'Englishness' that saw it as being more culturally hybrid and no longer white dominated. Kureishi became known through this early film work for exploring the complex and changing nature of both Asian and British identity against a background of economic crisis, unemployment, racism and social conflict. It is this sense of crisis brought about by the policies of the Thatcherite 1980s that I would argue provides the social and historical background to the *The Buddha of Suburbia*, published in 1990, but written between 1987 and 1989 (see Kaleta, 1998: 63). Its return to a more 'positively evaluated past world' (Tannock, 1995: 454) of the 1970s is a response to the perceived deficiencies of the1980s.

Such a concern with socially relevant issues helps to place Kureishi's work within the dominant realist discourse of British cinema and television. John Hill has argued that the realism of the two Frears/Kureishi films can be differentiated from the narrative and realist conventions of earlier British films like the working class realist cinema in the early 1960s (Hill, 1999: 216-218). These two films show a looser and more episodic narrative structure and a realism that is linked to a formal hybridity which Hill argues includes mixing genres, using a more stylised *mise en scène* and the use of symbolism. In thematic terms there is less evidence of the working class problems, characters, lifestyles and communities that had been present in the earlier 'new wave' films. These aspects have been replaced by a greater emphasis on middle class characters where class is just one element in relation to a diversity of gender, ethnic and sexual differences. Also, characters are no longer rooted in the fixed and stable communities of an earlier era, but operate in more fluid and shifting networks of personal relationships located in inner city and metropolitan areas of London. Many of these elements were to recur in the adaptation of his successful first novel, with the crucial difference that the story of *The Buddha of Suburbia* was set in the recent past of the 1970s.

## Generic Hybridity

In the institutional context of television such a realist/hybrid approach to issues of race and cultural identity was still in a minority and at the edge of the television mainstream in audience terms[5]. The adaptation of *The Buddha of Suburbia* put Kureishi's career closer to the centre of its dramatic traditions. The novel had a number of elements that made them an ideal vehicle for the 'quality' remit of a hard pressed public service organisation like the BBC. The critical success of his first novel[6] enhanced Kureishi's status and helped raise the possibilities of its future adaptation as a drama serial. The rarity of a successful British Asian author provided a commercial opportunity to market a 'quality' drama that dealt entertainingly with controversial issues for what would be a mainly white middle class audience. Like his earlier filmscripts the novel's social realism was interwoven with a generic hybridity. As a satire of metropolitan and suburban identities, lifestyles and sexual mores it can be viewed generically as a comic novel and as a journey of youthful initiation or *bildungsroman* (Hashmi, 1993: 26). Comedy is a very popular genre and there have been many examples in recent years of contemporary comic novels being successfully adapted as comedy dramas for television: Malcolm Bradbury (*The History Man*, BBC2, 1981), Michael Dobbs (*House of Cards*, BBC 1, 1990) David Lodge

(*Small World*, ITV, 1988; *Nice Work*, BBC 2,1989), Tom Sharpe (*Blott on the Landscape*, BBC 2, 1985, shown BBC 1, 1993); *Porterhouse Blue*, Channel 4, 1987) and Fay Weldon (*The Life and Loves of A She-Devil*, BBC 2, 1986).

In addition, to its ironic explorations of cultural identity the story is set within a pastiche of the period that is most evident in its references to the popular culture and music of the 1960s and 1970s (Kaleta, 1998: 82-3). These nostalgic elements led to it having some similarities to, but differences from, one of the most popular television drama genres of recent years, the 'classic serial' or literary historical adaptation. The growing commercial importance of nostalgia to the film and television industry is seen in the growth of 'heritage' dramas and films throughout the last 30 years. However, the 1980s saw big budget co-productions like Granada TV's *Brideshead Revisited* (1981) and *The Jewel in the Crown* (1984) come to be seen as that decade's epitome of 'quality' television (Brunsdon, 1990: 84-6). Their high production values and quality British acting focusing around the visual pleasures of recreating in detail the lifestyles of the upper-class and the aristocracy. A major criticism of such 'heritage' dramas has been that they present a nostalgic and uncritical representation of past aspects of British culture and society for a politically conservative era. However, Sarah Street has suggested that this is a limited reading as some form of social critique is present in many of these 'conservative' texts, and their nostalgic fascination with the past has a strong resonance with contemporary fears and anxieties (Street, 1997: 104). This past/present relationship also forms an important element of *The Buddha of Suburbia* which was a distinctive part of a growing trend for historical dramas to draw on a postmodern nostalgia for a more recent past, as well as focusing more on the lives of ordinary people. Some recent examples are mainstream dramas like *The Darling Buds of May* (YTV, 1991-93) and *Heartbeat* (YTV, 1992 – present), as well as more critical realist dramas like *Oranges Are Not the Only Fruit* (BBC2, 1990) and *Our Friends in the North* (BBC2, 1996). On its own, *The Buddha of Suburbia*'s comic pastiche of the 1970s might have allowed it to be seen as a drama without a serious social commentary. However, by comparison with Raj Revival dramas like *The Jewel in the Crown* or *The Far Pavilions* (ITV, 1984), its exploration of the 1970s offered a critical vision of the complexities of British and Asian cultural identity that was equally relevant to audiences in the 1990s.

## Literary Adaptation

At first hand it may appear that the aim of an adaptation is to reproduce all the elements of a novel. Hence, the reaction of many literary orientated critics to a film or television adaptation is to debate whether it has remained faithful to the 'spirit' of the novel. What often lies behind these discussions of 'fidelity' is an assumption that the literary text is superior to the adaptation. A reproduction of a taste hierarchy that still privileges the written word, be it a nineteenth-century novel or Shakespearian play, from its visual counterpart. This viewpoint is presented in an amended form in Bart Moore-Gilbert's recent exploration of *The Buddha of Suburbia* (see Moore-Gilbert, 1999: 274). He argues that the transfer of the novel to the screen has seen it remain faithful to its focus on issues of cultural identity and hybridity. Nevertheless, this literary approach still privileges the written word over the visual in its refusal to analyse the TV film, and reproduces a common-sense view of television as a transparent technology. It is a view that ignores how the formal codes and conventions of television help to mediate a text for an audience. In order to try and avoid such cultural assumptions John Ellis has argued that the real aim of any adaptation is to trade upon the memory of the novel, and the successful adaptation is the one that is able to replace the memory of the novel with its own (Ellis, 1982: 3). However, even the validity of this viewpoint has to be questioned as only a small proportion of an audience may have read an original novel or play especially when in the case of *The Buddha of Suburbia* it is adapted within only three years of its publication. Therefore the 'cultural memory' of the novel was still limited to a relatively small readership compared to the adaptation's reputed 5 million viewers (Moore-Gilbert, 1999: 274). As only a minority of the audience would have read the book by 1993, the television version would have undoubtedly played a dominant role in shaping the audience memory and of course encouraging many more people to read the novel.

The Kureishi/Michell version of *The Buddha of Suburbia* makes use of the narrative and realist conventions of dominant cinema. The four part serialisation uses a similar narrative structure to the novel with relatively minor modifications to the plot. The sexually transgressive desire of Karim Amir's father Haroon, for his 'friend' and mentor Eva Kay, initiates a plot that propels the hero Karim into a series of episodic adventures that ultimately leads to him leaving home as a result of the break-up of his parents marriage. The adaptation is not pushing at the boundaries of realism in the way that the Frears/Kureishi films do and it is therefore formally more conservative. The exception to this being the lack of plot resolution that occurs in the novel which is reproduced in the ambiguous ending of the adaptation (Kaleta, 1998: 113). However, Karim's ironic first person narration is not replaced in the television version by his off-screen commentary, but by a move to the more impersonal and 'authorless' mode of enunciation of the classic realist text (MacCabe, 1974: 10). This approach has two effects. It helps to reinforce the sense that Karim drifts with events and submits to the will of others until his later success as an actor gives him a direction in his life. This leads to a greater emphasis on the importance of visual codes and conventions through the use of the TV film's iconography and *mise en scène*. Its surface realism supports an 'empirical notion of truth' (MacCabe, 1974: 8), in that the spectator's knowledge about characters and their environment is delivered via what he or she observes through the omniscient 'eye' of the camera. Whilst this is a significant aspect of any mainstream adaptation it also involves varying degrees of inter-textuality in relation to the original novel. This is what Kerr terms its 'fidelity and authenticity' to the original text through the use of 'direct quotations from the novel in terms of both dialogue and decor (settings, costume, etc)' (Kerr, 1982: 13). The Kureishi/Michell adaptation is no exception to this generalisation as it draws from the novel's postmodern nostalgia for the period, as well as commissioning a musical score that is a 'dazzling pastiche of Seventies rock n'roll' (Thomas quoted in Kaleta, 1998: 107) from 1970s pop icon David Bowie. Whilst literary historical adaptations offer spectators a complex *mise en scène* they also involve 'repressed libidinal pleasures' that point to the gendered nature of spectatorship present within these dramas (Sonnet, 1999: 58). Such sexual politics are a crucial aspect of 'heritage' films and play an equally important role in *The Buddha of Suburbia's* representation of issues of race and identity.

## Race, Hybridity and Stereotyping

One of the areas of identity explored by Kureishi's novel is the question, 'what it means to be British?' Ian Chambers has argued that there are two versions of Britishness, one that '...is Anglo-centric, frequently conservative, backward-looking and increasingly located in a frozen and largely stereotyped idea of the national culture. The other is ex-centric, open-ended and multi-ethnic' (Chambers, 1989: 94). The former version is based on a colonial and imperial past that no longer exists in the form that it did. It has led Berthold Schoene to argue that this situation creates a problem of identity for the white English middle classes of the novel, in that '… Rushdie and Kureishi emphasize the lack of meaningful self-identification, the general post-Imperial vacuity and the intense cultural claustrophobia of contemporary England' (Schoene, 1998:113). In the novel, the middle class white characters search for a new self-identity is often expressed through an obsession with the exotic. In particular, the exotic as represented by a fetishisation of Indian culture that is a hangover from a colonial past. This manifests itself in their use of cultural stereotypes to represent their 'constructed, and deeply 'English', view of 'India'' (Chambers, 1989: 93). One example of such a character is Eva Kay, an ambitious but 'artistic' middle class woman trapped in a loveless marriage. Her sexual attraction to Karim's father, Haroon, is based not on the English side of his hybrid identity (he works as a low ranking civil servant), but on her perception of his ethnic exoticism. An exoticism that in his case is made up of several qualities, including his child-like and vulnerable nature, his Indianness, his interest in Buddhism and business potential as a guru. The exoticism of Haroon's new found identity also meets other English suburbanites needs for spiritual enlightenment. Haroon is a 'highly marketable commodity' (Schoene, 1998:114) as his visit to the Chislehurst home of a middle class English couple reveals that their house is filled with Indian antiquities and *objet d'art*, evidence of the fashionable interest in Asian ethnicity and exoticism in the 1970s. The couple's barefeet, bowed greetings and Indian attire lead Karim to remark to his father in the adaptation that, 'They know more about India than you do'. Despite Karim's perception of the inauthenticity of his father's new found spiritual identity as a guru, the stereotypical white perceptions of Haroon's exotic nature allow him to appear 'classless'. This is seen in his ability to mix socially with suburban whites from a higher middle class fraction than he comes from and, later on, to move in London's artistic and bohemian circles. His exoticism thus allowing him to cross class and ethnic boundaries and to also develop sexual relationships with white women.

## The Asian Male as Erotic Object

Like his father, Karim's hybrid identity is also constructed as an exotic. However, I would argue that the tele-adaptation intensifies this process through Karim becoming an object of the gaze. Visual pleasure is a major dimension of any film narrative, and the relationship between the camera/spectator's gaze and issues of patriarchal sexual identity has been well established by feminist film theorists like Laura Mulvey. However, Mulvey's original psychoanalytic focus on women as the objects of a masculine gaze ignored important areas like male objectification by a feminine gaze as well as the race and ethnicity of both characters and viewers. These elements have been more recently brought to the forefront of debates about the nature of spectatorial pleasures. This can be illustrated through an exploration of the complex inter-relationship between race, ethnicity, gender and sexuality in an early scene from episode one where we see Karim dressing in the privacy of his own bedroom. The scene is a short one in the novel, but is used in the adaptation as an early way of visually establishing Karim's hybrid identity. Karim becomes an object of the camera and the spectator's gaze, and in the process his cultural difference is constructed, like his father, through the stereotype of the exotic Other. This fits into a long established historical pattern of reducing the racial Other to an essential sexuality. His exoticisation becoming a source of voyeuristic and fetishistic pleasure for the audience. These elements are present through Karim's physical and cultural characteristics, in particular the objectification of his body and his flamboyant and narcissistic dress sense.

In the novel, Kureishi gives Karim the nickname of 'Creamy' which signifies not only his lighter skin colour but the biological rather than cultural origins of his hybridity. This is reinforced in the adaptation by the camera exploring the physical beauty of the actor Naveen Andrews[7]. Karim's male masquerade is an inter-textual reprise of a bedroom scene from the 1970s disco classic *Saturday Night Fever* (1977). Steve Neale comments that the Hollywood musical is the 'only genre where the male body has been unashamedly put on display in mainstream cinema in any consistent way' (Neale, p.18). The tele-adaptation shows a similar emphasis on Karim's 'feminization'. Both narcissistic and exhibitionistic traits are present as Karim, like Travolta before him, takes a great deal of trouble over his image in front of the mirror. Like Travolta, the camera gazes at Karim gazing at himself, and establishes him as a male object of desire. These 'feminine' traits help to contribute to Karim as having a less controlling and more passive masculinity. This is further reinforced by his 'in-between' social identity as an adolescent, in that he is not yet independent of the parental home.

The film's visual emphasis on his exotic 'otherness' also helps to mediate the novel's construction of him as an 'innocent' who is able to cross class, ethnic and sexual boundaries through the fascination he holds for a range of white Englishmen and women. In terms of spectator-text film theory Karim's fetishisation can be seen as reducing the potential threat his 'otherness' might potentially pose for a white audience by constructing him as a passive Anglo-Asian male. This very lack of threat suggests that Karim's appeal in the adaptation comes from the visual emphasis on his more 'feminised' and masochistic identity as constructed through his objectification, whereas, the novel's construction of him as an exotic innocent is juxtaposed with the knowing irony of his first person narration. An early example of this occurs when he first meets Eva Kay: 'When Eva moved, when she turned to me, she was a kind of human crop-sprayer, pumping out a plume of Oriental aroma … she looked me all over and said "Karim Amir, you are so exotic, so original! It's such a contribution! It's so you!" "Thank you, Mrs Kay. If I had more notice, I'd have dressed up" ' (Kureishi, 1990: 9). Karim's irony enables him to retain a distance from events and therefore a sense of his own power, even when he is being 'used or abused' by other Asian or white characters. However, the film text's use of an impersonal narration does not allow him this option and whilst the ironic dialogue from this scene is retained, Karim's commentary is lost. Thus Karim's 'feminine' identity is constructed in the film by a complex interaction of dialogue, acting, *mise en scène* and the degree to which he is or is not placed in control of the filmic gaze.

A further example of this complexity can be seen in the processes of disavowal that operate in relation to his objectification. Taboos against female and gay looking at the eroticized male body have become increasingly less common in the 1980s and 1990s, especially in pop music videos and TV advertising. However, it is a process that always threatens to destabilize the cultural norms of heterosexual masculinity. Hence, the significance of the process of disavowal which is used to both justify the voyeuristic and fetishistic gaze at the male body and to deny the 'taint' of homosexuality in a heterosexual culture (see Neale, 1993:19). There are various ways by which disavowal can be achieved but in this instance the eroticisation of Karim's body is disavowed by the use of a comedy of male display. Disavowal occuring when he strips to his underpants and then proceeds to comically powder his private parts with talcum powder. The homoerotic connotations of his objectification are disavowed by this visual joke whilst calling attention to his genitalia and his hidden (literally) desire for sexual adventure. A second example of comic disavowal occurs when the excessiveness of his identity is met downstairs by a disapproving stare from his father and disparaging comments from his mother: 'Don't show us up, Karim, You look like Danny La Rue'.[8] His mother's unconscious remarks hiding class and sexual taboos that are particularly revealing of the suburban need for conformity. They show a desire for both maintaining social status through keeping up appearances whilst at the same time revealing a fear of her son being perceived as homosexual. This is doubly ironic, as it is an aspect of Karim's sexuality that is later explored in detail through his sexual encounter and subsequent fraught relationship with Charlie Kay, the son of his father's mistress.

Karim's exotic appeal is tempered by the way that Kureishi shows how white characters try to make him feel inferior or are just plain ignorant about his background. Several examples drawn from the novel are used in episode one, for example, an Enoch Powell supporter calls him a 'wog', a 'blackie' and a 'nigger'; his history teacher calls him 'Pakistani Pete'; Charlie Kay believes that Karim meditates at breakfast, and a white girl, Helen, is surprised to hear that he comes from Bromley. Karim's ambivalence toward a culture of white English racism is expressed at the start of the novel through his ironic commentary: 'Englishman I am (though not proud of it)' (Kureishi, 1990: 3). Despite this verbal and visual stereotyping he still remains primarily a young Englishman whose Englishness in the adaptation is signified through a nostalgic pastiche of 1960s and 1970s pop culture. The bedroom scene being an example of what Simon Frith calls 'Bohemia in a bedroom: an alternative lifestyle practised at home' (Frith, 1997: 272), a cultural pick 'n'mix of posters (a Beatles Sgt Pepper's cut out), books (a Jack Kerouac novel), wall collage, LP records and clothing, all set to the tune of 'Get It On' sung by 1970s glam rock star Marc Bolan.

This early *mise en scène* also reveals how Karim's exoticism is constructed through an ironic use of period clothing. As a 1970s fashion victim he dons crushed velvet flared trousers that are excruciatingly tight, a flowery shirt, a mauve suede jacket, Cuban heels and a coloured scarf to act as a headband for holding his long black hair in place. The clothes not only add to his exoticism but as signifiers of the era help to enhance the 'authenticity' of the scene, their 'bad taste' being part of a sense of nostalgic irony that is part of a postmodern approach to the era. Leon Hunt has argued in his analysis of what he calls British low culture in the 1970s, that the nostalgia for the decade is double edged. In popular memory it is regarded as both a Golden Age, 'a signifier of innocence before a Thatcherite, style-obsessed Fall' (Hunt, 1998: 5), as well as an era of bad taste, a view that the 1970s, was not just stylistically, but politically unsound in its populist assimilation of cultural changes like 'permissiveness' and 'feminism'. Such a nostalgic irony about the gaucheness of the 1970s is based, Hunt argues, on a superior knowing relationship to the past (Hunt, 1998: 6).

## The Politics of Inter-racial Sexuality

If Karim is constructed by both film and novel as having a hybrid identity, it is worth remembering that the term 'hybridity' itself is problematic. The very term 'British Asian' suggests two cultures when neither are homogeneous and are in fact hybrids themselves made up of class, gender, sexual, religious, national and age differences. This hybridization of the hybrid points to the impossibility of essentialism in cultural identity and the importance of relating its shifting meanings to specific historical and social contexts. For example, in the nineteenth-century it was a term that was increasingly applied to cross-cultural relationships via Darwinian influenced theories of racial difference in an era of colonial exploitation. Robert Young has pointed out that these racial theories were themselves cultural in that they were imbued with white fears and anxieties about keeping the races apart, and points to the 'problematic of sexuality at the core of race and culture' (Young, 1995: 19). Such anxieties are the reverse side of an active sexual desire for the other that was expressed in the literature of the period which reveals 'the uncertain crossing and invasion of identities: whether of class and gender ... or culture and race' (Young, 1995: 2-3). An inter-racial sexual desire that is also at the centre of *The Buddha of Suburbia's* exploration of identity. There is an early suggestion in episode one that Karim is bi-sexual, reinforced by his lovemaking with his idol Charlie Kay, and later on Jamila his 'cousin' and Helen, a young white girl. Karim becomes infatuated with Eva Kay's charismatic son Charlie, an androgynous David Bowie look-a-like (played by Steven Mackintosh) who himself has ambitions to become a rock star and assumes various 'pop' identities in his increasingly desperate search for success. On a visit with his father to Eva's house, Karim follows Charlie's suggestion that they avoid the adults' spiritual explorations by retiring to his bedroom loft to listen to music. Unlike Karim's bedroom, Charlie's large loft suggests an informal but more solidly middle class bohemia. It is another visual pastiche of 1960s and 1970s youth culture whose *mise en scène* is this time supported by the diegetic music of Pink Floyd. It forms the background to Karim's masturbatory pleasuring of Charlie to orgasm, a representation of homoerotic and inter-racial lovemaking that not only points to the different power relationships between the two adolescents, but also avoids use of the earlier processes of disavowal.

If both Charlie and Karim are voyeuristic objects of the camera and our spectatorial gaze during their lovemaking, Karim's visit also involves him in becoming a controlling subject of the gaze. Before his lovemaking with Charlie he makes a trip to the bathroom which is accompanied by point-of-view shots of the walls which are covered with erotic Indian period prints of copulating couples. These exoticised simulacra not only arouse

Karim's curiousity, but they are suggestive of the erotic nature of Eva's middle class obsession with Indian culture. The *mise en scène* is prescient in that it is followed by Karim's startled voyeurism as he inadvertently becomes witness to his father's love-making with Eva. The use of his point-of-view helping to align us with Karim and making a strong contrast to the homophobic reaction of his father to his own later and accidental discovery of Karim and Charlie's lovemaking.

In the context of television this representation of homoerotic and inter-racial love-making for audiences is still extremely rare. The scene is an inter-textual reference back to the earlier controversial lovemaking between Johnny and Omar in the Frears/Kureishi film *My Beautiful Laundrette*. However, Karim's active 'gayness' is a passing phase in *The Buddha of Suburbia*. The novel/adaptation explores none of the gay scene as represented in the tele-adaptation of Armistead Maupin's *Tales of the City* (1993, Channel 4) set in 1970s San Francisco, or a more recent drama like *Queer as Folk* (1999, Channel 4), based in 1990s Manchester. What is interesting about the episode at Eva Kay's house, with its combination of irony, visual pastiche, popular music, sexual and spiritual explo-ration, is that it helps to both feed-off and reproduce the popular memory of the 1970s as a more innocent, less alienated and hedonistic period in relation to the more conserva-tive 1980s. We are left in no doubt as to the lingering on of the 'permissive' 1960s in the 1970s, a decade that the New Right and Mrs Thatcher continually condemned in their ideological support for a return to the values of the Victorian era.

## Conclusion

However, if the sexual libertarianism of this male utopia is all that there is to *The Buddha of Suburbia* then Kureishi's 'Golden Age' would be as one-dimensional and as monolithic as the Thatcherite view of the 1960s. The adaptation's visualisation of Karim's Anglo-Asian identity through racial and sexual stereotyping is in keeping with the novel's construction of him as exotic. However, this is far from Schoene's view of Karim in the novel as an 'ethnicity-free-no-man's-land' (Schoene, 1998:117) and also calls into question any evidence for Bhabha's notion of a Third Space. The idea of the Third Space has been seen as prematurely Utopian (Moore-Gilbert, 1999: 280) both in the context of Kureishi's novel and the historical legacy of racism left by the forces of colonialism. It is this tension between what Jussawalla has called the desire for being at 'home' with oneself and 'those who are considered natives' (Jussawalla, 1997: 28 ) that is at the heart of Kureishi's ambivalence towards a racist English culture. This desire to be at home with oneself and friends is expressed in Schoene's view that Karim's success as an actor sees him reach a kind of cosmopolitan Utopia by the end of the novel. This is mirrored at his final celebratory dinner party where Karim muses on both his past and future. Significantly, the adaptation changes the emphasis and sets this final scene on the evening of Mrs Thatcher's election victory in 1979 and intercuts black and white television news footage of the BBC's election night broadcast with the dinner party celebrations. These images of a victorious Margaret Thatcher and a defeated James Callaghan are juxtaposed with the immediate warmth and goodwill expressed by Karim and his multi-ethnic 'community' of friends. However, it is the final held close-up on Karim's face that, because of the historical references captured in the news footage, suggest a more uncertain and less Utopian future for Karim.

At the centre of Kureishi's postmodern nostalgia is what Wheeler has called an 'affective expression of the desire for community' (Wheeler, 1994: 94-5). It is a Utopian desire for a non-alienated state expressed I would argue at a time in the Thatcherite late 1980s when the sense of an all-inclusive and multi-ethnic community had been under attack from the policies of the New Right. Despite the Utopian limitations of the Third Space, the examples of cultural hybridity expressed in both novel and adaptation do have an historical and social basis, in that British Asian youth in the 1990s, like Karim Amir in the 1970s, have to negotiate their own cultural identities out of a complex and changing situation, the novel itself being a part of those wider social changes that have seen young Asians struggle to develop their own cultural identity. The rise of crossover musical hybrid forms like Bhangra-reggae-rap are proof of how some young Asians are reinterpreting their parents South Asian cultural identities in the context of British society (Sharma, 1996; 35-6). However, there is no sense here of identity being constructed out of a nostalgia for the past or a Utopian Third Space, anymore than is the case with the move to a more fundamentalist Muslim identity that is explored in Hanif Kureishi's latest film *My Son the Fanatic* (1997). Each of these identities, in their different ways, are responses to the culture of racism that still exists in British society in the 1990s.

## Notes

1   The term post-colonial has gained wide circulation in the Anglo-American academy. It has a complex and somewhat problematic history as a field of academic criticism and theory. The use of the term 'post' can imply the end of the period of colonialism (a somewhat contentious issue). It can refer to a range of criticism that explores the histories and subjectivities of societies that have been subordinated to European colonial power. Also, it includes a theoretical explanation of discursive practices that through the processes of migration and 'globalization' offer resistance to colonial or neo-colonial ideologies. It is this last sense that I am most concerned with in my exploration of *The Buddha of Suburbia*.

2   The term cultural hybridity has a variety of interpretations. I am using it here in relation to Homi K. Bhabha's concept of the Third Space where hybridity is more than just a fusion of two different identities but a constantly changing exploration of cultural identity. See Robert J.C.Young (1995) for a fuller explanation of the historical development of the concept of hybridity.

3   For the application of Bhabha's theories to Hanif Kureishi's novel *The Buddha of Suburbia* see Bernard Schoene (1998) and Bart Moore-Gilbert (1999).

4   See Sarita Malik's (1996) exploration of the role of Channel 4 in financing Black British films.

5   The audience figures for the first TV transmissions of *My Beautiful Laundrette* (1985) and *Sammy and Rosie Get Laid* (1987) were respectively 4,336 million (19.2.87) and 3,061 million (8.3.90). See Pym, J., *Film on Four, 1982/91: A Survey*, London, British Film Institute, 1992.

6   *The Buddha of Suburbia* won the Whitbread Award in 1991 for Best First Novel.

7   Naveen Andrews also plays a young Indian exotic in the BBC's two-part adaptation of Rumer Godden's novel *The Peacock Spring* (9.00 p.m., BBC 1, 1st January 1996). In the same vein as *The Jewel in the Crown* and set in New Delhi in 1959, the story recounts the doomed inter-racial love affair between Ravi, an under-gardener, and the daughter of his employer, an English diplomat.

8   These lines in the adaptation are taken unchanged from the novel.

# Bibliography

Bhabha, Homi K. 'The Commitment to Theory' in *The Location of Culture*, London, Routledge, 1994.

Brunsdon, C., 'Problems with quality', *Screen*, 31/1, 1990, pp 67-90.

Chambers, I., 'Narrative of Nationalism, "Being British"', *New Formations* no.7, 1989, pp 88-105.

Ellis, J., 'The Literary Adaptation- An Introduction', *Screen* 23/1, 1982, pp 3-5.

Frith, S., 'The Suburban Sensibility in British Rock and Pop' in Silverstone, R. (ed), *Visions of Suburbia*, London, Routledge, 1997.

Hall, S., 'New Ethnicities' in Kobena Mercer (ed), *Black Film: British Cinema*, ICA Document no.7, London, British Film Institute/ICA, 1989, pp 27-31.

Hashmi, A., 'Hanif Kureishi and the tradition of the novel', *Critical Survey* 5/1, 1993, pp 25-33

Hill, J., *British Cinema in the 1980s*, Oxford, Oxford University Press, 1999.

Hunt, L., *British Low Culture: From Safari Suits to Sexploitation*, London, Routledge, 1998.

Jussawalla, F., 'South Asian Diaspora Writers in Britain: "Home" versus "Hybridity"', in Kain, G., (ed), *Ideas of Home: Literature of Asian Migration*, East Lansing, Michigan State University Press, 1997.

Kaleta, Kenneth C., *Hanif Kureishi: Postcolonial Storyteller*, Austin, University of Texas Press, 1998.

Kerr, P., 'Classic Serials – To Be Continued', *Screen* 23/1, 1982, pp 6-19.

Kureishi, H., *The Buddha of Suburbia*, London, Faber and Faber, 1990.

MacCabe, C., 'Realism and the Cinema: Notes on Some Brechtian Theses', *Screen* 15/2, 1974, pp 7-27.

Malik, S., 'Beyond "The Cinema of Duty"? The Pleasures of Hybridity: Black British Film of the 1980s and 1990s', in Higson, A. (ed), *Dissolving Views: Key Writings on British Cinema*, London, Cassell, 1996.

Moore-Gilbert, B., 'Hanif Kureishi and the Politics of Cultural Identity', in Stokes, J. & Reading, A., (eds), *The Media in Britain: Current Debates and Developments*, , London, Macmillan Press, 1999.

Neale, S., 'Masculinity as Spectacle: Reflections on Men and Mainstream Cinema', *Screen* 24/6, 1983, reproduced in *Screening the Male: Exploring Masculinities in Hollywood Cinema*, Cohan, S. & Hark, I.R., (eds.), London, Routledge, 1993.

Ross, K., *Black and White Media: Black Images in Popular Film and Television*, Cambridge, Polity Press, 1996.

Schoene, B., 'Herald of Hybridity: The emancipation of difference in Hanif Kureishi's *The Buddha of Sububia*', *International Journal of Cultural Studies* 1/1, 1998, pp 109-127.

Sharma, S., 'Noisy Asians or 'Asian Noise'?', in *Dis-Orienting Rhythms: The Politics of the New Asian Dance Music*, in Sharma, S. Hutynk, J., and Sharma, A. (eds), London, Zed Books, 1996.

Sonnet, E., 'From *Emma* to *Clueless*', in Cartmell, D. and Whelehan, I. (eds) *Adaptations From Text to Screeen, Screen to Text*, London, Routledge, 1999.

Shohat, E./Stam, R., *Unthinking Eurocentrism*, London, Routledge, 1994.

Street, S., *British National Cinema*, London, Routledge, 1997.

Tannock, S., 'Nostalgia Critique', *Cultural Studies* 9/3, 1995, pp 453-464.

Torrey Barber, S., 'Insurmountable Difficulties and Moments of Ecstasy: Crossing Class, Ethnic and Sexual Barriers in the Films of Stephen Frears' in Friedman, L. (ed.) *British Cinema and Thatcherism*, London, UCL Press, 1993.

Wheeler, W., 'Nostalgia Isn't Nasty: The Postmodernising of Parliamentary Democracy' in Perryman, M., (ed) *Altered States: Postmodernism, Politics, Culture*, London, Lawrence & Wishart, 1994.

Young, R. J. C., *Colonial Desire: Hybridity in Theory, Culture and Race*, London, Routledge, 1995.

# The Grotesque and the Ideal: Representations of Ireland and the Irish in Popular Comedy Programmes on British TV

Margaret Llewellyn-Jones

## Introduction

This chapter considers the representation of Ireland and the Irish as seen on British TV, through comparative analysis of the successful comedy series *Ballykissangel* and *Father Ted*. Discussion is rooted in the post-colonial cultural context, exploring the extent to which genre and ideology are inevitably linked to an absence of history through the use of stereotypes of Irish identity, especially in the context of gender and religion. The romanticising effect of landscape and tourism on TV representation is contrasted with the function of the grotesque – a key element in Irish literature – as a potential agent of critical realism. Brief reference to other genres is related to the notion that hybridity and fluidity are apposite for both reading and creating new TV texts and post-colonial identities.

This essay will explore issues associated with the representation of Ireland and the Irish in popular programmes shown on British TV, through detailed reference to the light comedy series *Ballykissangel* (shown on BBC 1), and the situation comedy *Father Ted* (originally shown on Channel 4), with more general reference to other fictional programmes set in Ireland. These two popular programmes have been chosen because they may resonate differently with certain long-standing prejudices in the British viewer, whilst conveniently ignoring key aspects of the history and complexities of the political situation on both sides of the border, before and during the ebb and flow of the faltering Peace Process. [1] A comparison of these representations will be framed by the significance of the post-colonial cultural context, and related to concerns about the relationship of form and genre to ideology which may affect readings of such texts. Theoretical references will allude to the differences between realism and the post-modern in terms of the function of time and space as well as Bakhtin's notion of carnival and the grotesque.

Both programmes are rooted positively in Ireland through aspects of the production processes, location and performers. *Ballykissangel* was initially directed by Richard Standeven, made by the company Ballykea Productions, commissioned by BBC Northern Ireland and World Productions, produced originally by the late Joy Lale, written by Kieran Prendiville with Graham Frake as Director of Photography. Four series have been shown on BBC1 from 11 February 1996–17 March 1996; 05 January 1997–23 February 1997; 01 March 1998–01 May 1998; 20 September 1998–06 December 1998. Some episodes are now available on video. Significantly the series was shown on Sunday evenings, in prime family viewing time, after *Songs of Praise* and the *Antiques Roadshow*. The main roles included Father Clifford, played by the British actor Stephen Tompkinson from the Channel 4 series *Drop the Dead Donkey* (from 09 August 1990) Assumpta Fitzgerald played by Dervla Kirwin known in England for her role in the early runs of the BBC1 series *Goodnight Sweetheart* (from 18 November 1990), and the village entrepreneur Brain Quigley played by the late Tony Doyle, an internationally known Irish theatre and screen actor who died early in 2000. Although the series was produced in Belfast, Northern Ireland, it is set in the rural South, and heralded by Mark Lawson (*The Guardian* 21 March 1996) as, like *Father Ted* 'a child of the ceasefire... product of a new order'. The first series of *Ballykissangel*, despite initial lack of interest from RadioTelefeis Eireann, was shown in the republic from May 1998. *Father Ted* was created by Irish writers Graham Linehan and Arthur Mathews, beginning as a one hour spoof documentary, which they then sent to the independent company HatTrick Productions in England, and thence to Channel 4. According to an interview (*Film West* Summer 1997 pp38-40), the writers were able to join in the the casting process, and suggested Dermot Morgan, Ardal O'Hanlon, and Frank Kelly for the main roles Father Ted Crilly, Father Dougal Maguire and Father Jack Hackett. In its first showing, Channel 4 placed *Father Ted* on Friday evenings betwen the American hits *Cybill* and *Roseanne*. The three series were originally shown 21 April 1995–26 May 1995, 8 March 1996–10 May 1996, 13 March 1998–01 May 1998, the second and third series on Friday evenings. Some have been repeated and are now available on video. Gaining a cult following, the series won a raft of awards including a BAFTA as best British Comedy, an Indie for Best Light Entertainment Programme, a Writer's Guild Best Situation Comedy, an Emmy Nomination for the 1996 Christmas Special, as well as viewers' votes for the best home-grown sit com. BAFTA also gave individual Awards as best Newcomer 1995, best TV comedy actor 1996, and best TV comedy actress to Ardal O'Hanlon, Dermot Morgan and Pauline McLynn, as Mrs Doyle the housekeeper, respectively. The 1999 BAFTAs awarded both Dermot Morgan posthumously for his performance, and the programme as best situation comedy. Oddly, the source of the greatest number of catch phrases, Frank Kelly as the outrageous Father Jack Hackett, did not get an individual award. Prompted by these successes, RTE eventually bought rights for showing this programme, too.

However, whatever the conditions of production and the input of both Irish and British elements, potential readings of both selected programmes, perhaps linked to television genre expectations, still have traces of old stereotyping. In view of Professor Mary Hickman's extensive recent research on anti-Irish racism in Britain,[2] seemingly such attitudes die hard. It is debateable whether the ethnicity of a joker ameliorates the dubious quality of a representation or a joke in this context. Nevertheless, as Lawson's article 'The Grins of the Father' (*The Guardian* 21 March 1996) pointed out, both programmes have been unexpectedly popular, with *Ballykissangel*'s viewing figures for the first series attracting up to 15 million viewers, ranking it as one of the BBC's most popular dramas ever, and confirming the results of focus group test viewings as the most positive market research response. The programme was financed due to John Birt's regional quota system. *Father Ted's* popularity is revealed not only in articles written in Ireland, but in the wide coverage of Dermot Morgan's untimely death, which postponed the showing of the third series in January 1998 for a week.

Roy Foster's *From Paddy to Mr. Punch* (1995) and Terry Eagleton's *Heathcliff and the Great Hunger* (1997) explore the historic roots of the British conception of the Irish as 'Other'. The former cites examples of the evolution of literary and often grotesque cartoon examples of the stereotype, which embody the kind of 'Stage Irishman' that, during the Literary Revival, the National Theatre Manifesto dedicated itself to defeating.[3] Eagleton also acknowledges the complexity of the colonial and postcolonial relationship between Britain and:

> an island ... unsettlingly close to hand ... it is not with Ireland simply a question of some unscrutable Other, as an increasingly stereotyped discourse of stereotyping would have it; it is rather a conundrum of difference and identity in which the British can never decide wheher the Irish are their antithesis or mirror image, partner or parasite abortive offspring or sympathetic sibling (1997 p127).

Geographical closeness thus intensifies the ambiguity of a relationship which, despite the passing of the notorious 1950s letting notices 'No blacks No animals No Irish' and the current prevalence in Britain of themed pseudo-Irish pubs, is still uncertain, partly due to a general lack of historical understanding as well as previous inflammatory reporting of the Troubles.

A more recent example of crude representation, which caused uproar even in the British tabloid press, was the visit of the popular soap opera *EastEnders* (BBC 1) to Ireland, when the plot line showed Pauline Fowler tracing an illegitimate half sister whose existence had only been revealed after her mother Lou Beale's death. The characterisation of the rural Irish relatives fulfilled the basic stereotypes, including large families, brow-beaten yet stoical mothers, peasant stupidity and trickery, and drunken men. This 'curse of Paddywhackery in soaps' produced 100 calls to RTE by lunchtime, and the chairman of the Irish Tourist Board condemned the soap for 'its negative image of Irish hospitality' (*The Guardian* 26 September 1997). The shallow approach was not entirely redeemed when two of the characters joined the more permanent cast in Albert Square, a niece predictably named Mary, and Conor, her feckless, usually absentee, father. At the present time of continued negotiations about the Northern Ireland situation including the Good Friday Agrement, and the expansion of the 'Celtic Tiger' economy of the Republic partly fuelled by EEC support, Irish identity is at an especially fluid moment, when such retrograde representations are especially inept, because they are fixed in the past.

> Cultural identity ... is a matter of 'becoming' not being. It belongs to the future as much as to the past. It is not something which already exists, transcending place, time, history, culture. Cultural identities have histories ... but ... they undergo constant transformation

... they are subject to the continuous 'play' of history, culture and power (Stuart Hall, quoted in (ed) Wayne 1998 p106).

As identity, in terms of gender or culture, can also be seen as performative, it is significant that Gilbert and Tompkins (1998 p12) in their analysis of post-colonial theatre performance prioritise acts which respond to imperialist experience especially in terms of gender and racial identity, acts which continue or regenerate colonised communities through deploying theatricalised cultural practices such as ritual or carnival, and acts which use the stage space and the performing body as sites of resistance. Clearly the use of camera and editing techniques in creating a TV text, and the nature of television spectatorship – which differs from theatre or film both in its domestic context and its glance rather than the gaze at the screen – does not provide a exact overlap with the theatre medium. Nevertheless, the relationship of *Ballykissangel* and *Father Ted* to dramatic form and strategies have ideological undertones significant for the post-colonial context and its link with gender representation. This essay suggests that in different ways both these TV dramas persist up to a point in constructing the Irish as 'Other', because post-colonial power relations are exnominated through dramatic forms which push history as it were outside the four walls of the domestic setting, or beyond the tourist-beckoning landscape. As Eagleton's (1995) discussion of Raymond Williams account of the space of naturalism implies:

> history ... is always offstage. The forces which shape these man and women are thus condemned by the dramatic form itself to remain invisible and opaque (p313).

Exnomination is particularly facilitated through television discourse, it is the process through which 'discursive power is hidden', and which:

> masks the political origins of discourse, and thus masks class, gender racial and other differences in society. It establishes *its* sense of the real as the *common sense* ... and thus invites the subordinate subcultures to make sense of the world ... through the dominant, exnominated discourse (Fiske 1987 p43).

Paradoxically, therefore, the absence of historical awareness allows the British viewer to enjoy bland representations of Ireland, whilst also facilitating the longevity of stereotypes which are rooted in colonial relationships.

Eagleton (1995 p9) has suggested that for the British, Ireland operates as if it were an unconscious site – 'raw, turbulent, destructive' but also 'a locus of play, pleasure, fantasy, a blessed release from the tyranny of the English reality principle'. The rural setting of *Ballykissangel* particular embodies this escapist approach, which is heightened by the opening and finishing credits in which a childlike, idealised cartoon-style picture of the eponymous village within the Irish landscape fades into and out of shots of genuine landscape. Throughout the episodes, scenery is often foregrounded for its pleasureable self. However, ever since the on-stage language of the Celtic Revival playwrights evoked off-stage landscape, from John Millington Synge's *The Shadow of the Glen* (1902), to Sebastian Barry's *The Steward of Christendom* (1995); the Irish landscape has also been a crucial signifier of Irish nationalism, through for example Padraic Pearse's role in the 'process of the recreation of nature as culture' (O'Toole 1994 p45). Further, and paradox-ically, as Luke Gibbons indicates, scenery has also been used to evade crucial issues:

> Landscape has tended to play a leading role in Irish cinema, often upstaging both the main characters and narrative themes in the construction of Ireland on the screen ((ed) Rockett *et al* 1988 p283) .

In addition to the voyeuristic pleasures the rural scene has for metropolitan viewers, it also enhances the notion of Ireland as tourist venue – through location visits to Avoca, a village in County Wicklow where the shooting of the series takes place. The village had suffered chronic unemployment since the closing of local copper mines in the early 1980s, and now has a minor economic boom in craft shops and tourist accommodation. This curious slippage between fact and soap fiction was compounded by a BBC *Songs of Praise* programme, publicised as from Ballykissangel and broadcast from the village church of St Patrick's and St Mary's, which is shown as St Joseph's in the series. (*The Guardian* 20 may 1996). TV location tourism – seen also in Britain as in trips to *The Last of the Summer Wine Country* or *Heartbeat Country* ad nauseam – blurs fact and fiction more insidiously when it overlooks history or re-interprets it for tourism. Father Dan Breen, the real parish priest, is reported in the above article as saying that his anxieties about the persistent presence of the TV team were alleviated when he was '...assured that it was all going to be whimsical and harmless, and that has proved to be true'. However, as Fintan O'Toole rightly complains about the Disneyfication of Ireland:

> The grand narrative of Irish Nationalist history has been destroyed, leaving a gap for the pop images to fill, not merely for the tourist but for the native as well. In the process, the real relationship of history and geography, the real narrative of the landscape is occluded (1994 p41).

The front credits of *Father Ted* show Craggy Island, pan across a desolate shore, stone-walled fields, and then from above look down on the isolation of the large square Georgian-style presbytery, which links to the theme of narrowness and entrapment fore-grounded throughout the series. Sometimes mid-episode shots show the darkness surrounding the presbytery at night. The end credits usually show the same presbytery scene but often include comic elements from the episode, such as the gush of water up from a remote drain into a visiting bishop's skirts. The fictional site of Craggy Island itself is deliberately kept flexible by the writers, Linehan has said 'the island grows or shrinks according to what we are doing' (*Time Out* 26 February 1996 – 6 March 1996). Even where episodes show the magnificence of the rural landscape, it is often undercut through the comedy, for example in the 'Hell – of a Caravan Holiday' (08 March 1996), where wet weather and irrepressible youth club hikers with another priest merely inten-sify the impossibility of escape. Location shots, which were fed on screen into the usually live recording of the episodes before an audience in a London Studio, generally came from Ennistymon in County Clare where locals participated as extras.

# Representations of Ireland and the Irish in Popular Comedy on British TV

Although *Ballykissangel* and *Father Ted* have some similarities, such as the rural location, formal and strategic differences make the former more conservative ideologically than the latter. Both are examples of the increased blurring of generic boundaries which has marked the evolution of television drama. Generically, *Ballykissangel* is a light comedy series with elements of soap opera; several lines of through narrative operate across episodes within each of the series, whilst within each episode a particular situation or incident is also explored, rather more as it might be in a situation comedy. In the first three series, the major hermeneutic link was the on/off potential sexual relationship between the English priest Father Clifford, and Assumpta Fitzgerald who ran the local bar. Soap opera elements can be related to the melodrama tradition, but in this context are also be underpinned by the strength of the Irish nineteenth century cultural tradition of theatrical melodrama, as for example the works of Boucicault. To a limited extent therefore it may be said that there is a trace of the post-colonial quality of re-working an indigenous art form. However, Gledhill has commented that melodrama may be:

> judged as an ideological construction – in which class divisions and struggle are dissolved, displaced by compensatory wants more easily satifiable by the capitalist culture industries (1987 pp36-7).

The glossing over of potential problems in feel-good location-based genres is thus scarcely radical. In contrast, *Father Ted* has a situation comedy structure, in which each episode is usually resolved, continuity being maintained by the central handful of characters in the Craggy Island presbytery, though extra characters may be introduced according to the needs of each episode. The form, despite this apparently mimetic setting and generally linear drive, may contain deconstructive aspects which are sometimes similar to Belsey's (1980) definition of the interrogative text, or which may seem excessively carnivalesque grotesque eruptions. These also, in some episodes, have an almost post-modern intertextual quality, and have strong links with the traditional aspect of the Irish grotesque, a point which will be developed further below in terms of post-colonialism. The style of dialogue is also closely related to stand-up comedy, an area of satirical performance in which Dermot Morgan and Ardal O'Hanlon have been very successful, and perhaps echo the oral tradition of the *seannachie*, the Irish story-teller. Where *Songs of Praise* blurred the fictive and the real Ballykissangel, the publication of volumes of *The Craggy Island Parish Magazine* mischievously aims for deconstructive effect, which interestingly also often debunks the colonising adventures of priests in other foreign contexts through grotesque exaggeration.

Although in terms of linear narrative plotting, the form of both TV texts has elements akin to classic realism as defined by Catherine Belsey (1980) and Colin MacCabe (in (ed) Bennet *et al* 1981), Robin Nelson's (1997) useful refinement of their terms would place *Ballykissangel* as an example of 'formulaic realism' similar to *Heartbeat*, whilst the latter has some elements of 'critical realism'. Quoting a line from Stoppard's *Rosencrantz and Guildernstern are Dead* (1966) to show how human beings live in an eliptical, blurred relationship to 'truth', Nelson is concerned with the nature of referentiality and ways in which the viewer may be nudged into re-focusing his/her perceptions of both world and representation:

> Formulaic realism is conducive to the permanent blur; critical realism is the agent of the grotesque' (Nelson 1997 p121).

That the whimsicality of *Ballykissangel* promotes a gentle smile remote from the rigours of history, though with some allusion to current economics, is borne out by the headline: 'No Bombs or Blarney in Gentle Irish comedy' (*Time Out*, 04 February 1996). In contrast much of the laughter provoked by *Father Ted* has the kind of dianoetic quality claimed for traditional forms of Irish humour, both macabre and grotesque, by Mercier (1962 pp48-9). In linking the former with terror and the fear of death, he associates the grotesque with a dread of the mysteries of reproduction. In this sense at least the latter programme is more in critically in touch with cultural history, and especially the role of the Roman Catholic church, through its re-working of the grotesque through absurdist satire and farce.

The macabre and grotesque humour which Mercier claims help humans 'to accept death and belittle life' fuels the programme's approach to the Fathers' characters and situation. Mercier's analysis of the evolution of the Irish comic tradition traces its development from the folk tradition, Gaelic comic literature, through Swift, the Literary Revival to Joyce and Beckett. He considers any 'archaizing movement is apt to beget a comic revival', and thus considers the persistence of grotesque and macabre humour crucial to Irish culture. In this sense *Father Ted* draws much more strongly upon the indigenous culture from within, whereas *Ballykissangel* seems to reflect its surface, re-presented as a commodity for those outside. The grotesque functions as a tool of critical realism through characterisations, formal and plot elements in *Father Ted*. Father Jack is the most obvious manifestation of the bodily grotesque. His gurning face, wild hair, bizarre postures and catch-phrase cries 'Feck!' 'Arse!' 'Drink!' 'Girls!' evoke not only areas taboo for the priesthood, but re-iterate the bodily boundaries celebrated by the carnivalesque grotesque as defined by Bakhtin ((ed) Morris 1994). His make-up, designed by Christine Cant, takes two hours to apply, and emphasises the apparant drooling of saliva, and droppings of ear wax which heighten this excessive yet abject effect.[4] The other two priests are represented subversively but differently, Ted, generally a well-meaning mediator, sometimes gets into overblown schemes; there may be financial irregularities in the past which he fears may come to light; Dougal is a holy fool, ignorant of even basic religious beliefs. Their fearful attitudes to sex and reproduction were clearly shown for example in the Christmas show (24 December 1996, rptd 24 December 1998) in which they and a further group of priests were trapped in the ladies lingerie department in town. Jungle sound effects and a series of comically erotic poses assumed by the female store dummies, around which the fearful chain of priests crept low like explorers to escape observation, underlined both priestly naivete and hypocritial double-think, since despite their embarrassment, they had, like naughty schoolchildren, made detailed observation of the underwear size and styling. Mrs Doyle, despite her cups of tea and unpleasantly mixed sandwiches, is a grotesque distortion of the idealised, stoically suffering, Irish woman celebrated by de Valera in the Constitution. [5] Her frequent brushes with disaster and consequent wailing are suggestive of the banshee, the grotesque old women whose cries are associated with death. The three priests could also be read as a satirical reverse gender echo of other archetypal roles of Irish womanhood, as seen for example in Tom Murphy's theatre play *Bailegengaire* (1985), with Father Jack as a whore, Ted as a mother, and Dougal as a virgin. Terror of death provoking macabre and grotesque humour is strong in an episode 'Grant Unto Him Eternal Rest' (26 May 1995) where an apparently dead Father Jack is laid out in the church, similar to J. M. Synge's *Shadow of the Glen* (1902). As in the canonical play, the corpse eventually rises – here as the effects of the boot polish he has drunk have now worn off.

Representation of gender in Ireland is inevitably linked to post-colonial factors especially as in the ambivalent iconography of the young Cathleen Ni Houlihan and the suffering old woman the Sean Bhean Bhoct. [6] The emasculation of the previously colonised male and the association of nationalism with beautiful but domesticated women is often a source of laughter in *Father Ted*, but paradoxically so is the attempt to challenge this stereotype. Feminist critics have pointed out the persistence from the late nineteenth and early twentieth century in Ireland the idea that:

> The image of the West stands at the centre of a web of discourses or racial and culturaliidentity, femininity, sexuality and landscape which were being used in attempts to securecultural identity and political freedom (Nash quoted in (eds) Stern-Gillet *et al* 1996 p180).

This symbolic and asexual idealisation of woman as Virgin/Mother in contrast to the Magdalen figure, combined with economic and social factors which have in the past either propelled men into working in exile or prolonged bachelorhood at home, has also affected notions of masculinity which reverberate in both programmes under scrutiny. The thwarted masculinity shown through clerical grotesqueries in *Father Ted* underlies the representation of the secular males in *Ballykissangel*. Declan Kiberd has written of the origins of this problem:

> The Irish father was a defeated and emasculated man. If successful, he lived out his life in a posture of provincial dependency as a policeman or a bureaucrat or a petty official in an oppressive and despised colonial administration. If unsuccessful he retreated into a vicious cycle of alcoholism and unemployment. In the home the mother often usurped his potential function as provider ... just as the priest tended to usurp his potentialrole as a spiritual leader (in (ed) Kenneally 1992 p132).

In *Ballykissangel* this masculine weakness is seen most strongly in Ambrose, the policeman, whose vacillation about personal matters is in direct opposition to his petty, professional bullying of fellow villagers. In *Father Ted* it is linked to the effects of priestly celibacy and the role's other restrictions which are intensified by isolation, and against which the body rebels.

The virtual incarceration of the priests within the confines of the Craggy Island presbytery has echoes of canonical existential European texts by Kafka, Sartre, and Beckett, where hell is both other people and the failure to communicate ideas. Across the episodes grotesque and excessive intrusions playfully satirise the power and ritual of the Catholic Church, as for example in the episode about the Holy Stone of Clonricket (*Tentacles of Doom* 22 March 1996), and the Christmas episode (24 December 1996) about Ted winning the Golden Award for the best priest in Ireland. In the first instance, a quick intercut scene of the high-ranking clerics in Rome showed them deciding with a 'whatever' and a careless wave of the hand to dedicate the stone. In the second, several shots of a highly coloured gold, purple and blue scenario showed a group of such clerics feverishly playing rock music were interspersed within the 'straighter' elements of narrative. Metatheatrical touches include the priests watching TV advertisements for a 'Priest Chatline'. Priestly excesses not only demonstrate a subversive carnivalesque quality expressed through the body, but also provide a critical dimension to those aspects which seem loosely realistic, but paradoxically are postmodern in their overlaying of different spaces and times. As Foucault suggests:

> We are in the epoch of simultaneity: we are in the epoch of juxtaposition, the epoch the near and the far, of the side by side, of the dispersed (1986 p22).

The collage-like juxtaposition of the quasi-real and the ridiculous nudges the viewer into questioning the contradictions.

The episode 'Rock a Hula Ted' (19 April 1996, second series) features several significant characteristics, including deconstructive attitudes to gender, religious institutions, tourism and the media. It begins with Ted and Dougal watching on TV a feminist rockstar, not unlike Sinead O'Connor, criticising religion, and singing 'Big men in frocks tell us what to do', accompanied by a miming woman. This metatheatrical aspect manages to deconstruct both the ridiculously overstated attitudes of the star, and the puzzlement of the priests over radical feminism and sexism. Apparently she is intending to create in Craggy Island a spiritual haven free of sexual and religious intolerance – a prospect which horrifies the priests. Their lack of awareness is underlined as Mrs Doyle, staggering under a builder's hod, is expected to make tea after a hard day's labour, and does not understand what sexism is. Further, Ted is delighted when asked to judge 'The Lovely Girls Contest' by a fellow priest whose general frustrations make him break furniture at random, wanting Ted to get the winner to wear a dress made by his mammy for the celebration dinner. The ambivalent Madonna /Whore polarity is variously underlined, as when Dougal, seen in close-up reading a magazine showing the rockstar with the mysterious label 'Clit Power', is given advice by Ted about dealing with women. At the Lovely Girls contest, the inept non-clerical males gaze giggling at the candidates, perhaps an allusion to the actual persisting problem of rural late marriage and the preponderance of bachelors. The contest which includes graceful walking, making thin sandwiches of an appropriate size, as well as the usual inaccurate interviews satirises the female ideal. Intercut with the contest narrative, is the rockstar's visit to the presbytery – where Dougal invites her to undo her bra so that she is more comfortable – her raptures over the cultural implications of the old furniture, faded decor, and iconic pictures culminate in her claim that this house fits her requirements exactly. Thus Ted's return, rejoicing in the prospect of a free dinner with the contest winner, is ruined because Dougal, faithful to the letter of his advice, has given the rockstar the presbytery. An unusually groomed Father Jack is seen, still at the contest, as the unlikely centre of a circle of admiring beauties. Eventually, the rockstar returns the presbytery to the priests on the understanding that Mrs Doyle is given one evening off per week. A celebratory 'sisterly' dinner unites the three female archetypes within one frame – the Lovely Winner (Madonna) the singer (Magdalen), and the grotesque banshee, Mrs Doyle. Unfortunately, the latter is shown as wrestling incapably with chopsticks, so any potential pro-feminist point is undercut by this mockery of rural ineptitude. Indeed, the superficiality of the rockstar's attitudes to landscape, religion and cultural signifiers is deconstructed throughout, thus suggesting implicitly that her apparently feminist position is only skin-deep, and part of her publicity. The final credits reveal the priests desparately throwing items around in the kitchen in their feeble attempt to get themselves a meal for once. The episode's satirical bite therefore is limited, since although it attacks aspects of dominant institutions associated with representations of Irish identity, including the Catholic Church, and ideas of remote Irish landscape as a spiritual haven; the episode's attitudes to gender are more ambivalent, and verging in the depiction of Mrs Doyle on the misogynistic.

Analysis of one typical *Ballykissangel* episode (25 February 1996, no 4, first series) indicates how far the series differs from *Father Ted* in attitudes to gender, whilst sharing some post-colonial elements. Within a linear narrative, the comedy is based on character, and primarily concerned with personal relationships. Combined with lingering visual pleasures associated with the rural setting these qualities suggest that this might be considered as a feminine TV genre. The range of character types, although indicative of rural values as opposed to urban disenchantment, does not draw particularly on folk tales, but provides stock community roles, typical for example of those found in Patrick Kavanagh's novel *Tarry Flynn* (1948), which was adapted for the stage by the Abbey Theatre, Dublin, in 1998. These roles include Eamonn, an elderly eccentric farmer; Brendan a school teacher; Siobhan a (female) farmer with veterinary skills; Liam a rather dim employee of Quigley, the local entrepreneur and builder whose daughter Niamh is on the verge of marriage to Ambrose, the over-enthusiastic but ineffectual local policeman, and a senior priest from another parish. These locals all exhibit aspects of Irish 'Otherness', which are in contrast to the puzzled but good-hearted English priest, Father Clifford, who is something of an innocent in this Eden. The main female character, Assumpta, is feisty, exhibiting more 'modern' almost feminist qualities in running the pub which is the social focus, and skirmishing with the priest. Significantly, these untraditional qualities were eventually punished – almost in nineteenth-century melodrama tradition – when she was written out via electrocution in the third series. Unusually, in one episode, a visiting character, Quigley's old flame, a professional academic working for the EEC, provides a link with the 'real' world of economics and regulations.

The episode 'Live in My Heart and Pay No Rent' (25 February 1996) has three major interwoven plot strands. Firstly, to the annoyance of his daughter Niamh, Quigley, a widower, has received a message that an old flame whom he rejected to marry his wife, intends to meet him on the local mountain-top where they used to do their courting. Secondly, because Ambrose has escaped being crushed to death by a falling holy statue outside the church, he decides he has a vocation and cannot marry Niamh. She goes ahead with 'hardly a wedding reception' in the pub. Thirdly, Eamonn is concerned that spy satellites will reveal the scarceity of his sheep to the EEC watchdogs, and thus reduce his EEC farming subsidy. A further minor detail is that Assumpta has laid off the (unnamed) draught Guinness, to the indignation of regulars who consider it is their 'cultural inheritance'.

All these initial difficulties, indicative of aspects associated with Irish identity, are satisfactorily solved through a triumph of contemporary action over traditional blarney. Quigley's assignation with his old flame is framed by the romance of nostalgia and landscape, but she is happily married and working for the EEC. Ambrose's mistaken vocation provides a slightly satirical opportunity which contrasts the senior Irish priest's wish to ordain him to offset falling clerical numbers, with the English priest's attempt to persuade him back into marriage through adages like 'A man who fears love fears life', and a lie about the saintly status of the falling statue. Eamon's anxiety about the sheep is solved by a painted, wooden flock with which he intends to the trick the EEC monitors. On the other hand, the postponed wedding and premature reception provide ample opportunity for evidence of 'feelgood' traditional community support – drink and dancing to nostalgic music. Further, the success of the reception encourages the Pub's suppliers to provide free draught Guinness until the tourist season starts, and prompts the agnostic Assumpta to donate a large sum to the church roof repairs. This unexpected gift underlines the sexual tension between herself and the priest which is manifest in shot-reverse shots as they exchange either witticisms or marked silences, with background music 'True Love Waits'. Throughout, there is evidence of sisterly support for Niamh from the other women, but the emphasis on marriage and the hermeneutic tease of unspoken priestly love offsets the marginally progressive elements within this and other episodes.

Limited space allows only for brief contrast between these two comedy texts and other recent TV drama representations of Ireland. Novel adaptations centred on problematic personal relationships in a rural environment rather than historical events have been popular recently. For example *Falling for a Dancer* from Dierdre Purcell's novel set in the 1930s, commissioned by BBC Northern Ireland, sponsored by Bord Scannan na hEireann, and produced by Parallel Films in association with Mayfair International was assisted by the European Script Initiative Fund for Media programmes. The serial shown on BBC 1 in 1998 was pre-sold to Australian Network 7, and to Carlton Home Entertainment for UK sales. *Amongst Women* based on John McGahern's novel set in the 1950s was shown on BBC 2 in 1998, and developed by BBC Northern Ireland and by Parallel Films in association with RTE and also the Irish Film Board. Both provide a starker view of the problems of women within a rural context, in the relatively recent past. The latter, reviewed favourably in *Film Ireland* (no.65 June 1998) and *Film West* (no 33 July 1998), starred Tony Doyle as a harsh widowed patriarch with four daughters and two sons, showed how, even after his remarriage, religious oppression and limited economic and educational opportunities, tragically restricted both his sons and daughters. The former explored the fate of a woman with an illegitimate daughter, struggling to survive in beautiful but harsh landscape and even harsher moral attitudes, though eventually finding a partner, thus re-instating dominant values in closure.

On the other hand, the series *The Hanging Gale*, produced by Little Bird and featuring the Liverpool McGann brothers, seen on BBC1 on Sunday evenings in May – June 1995 was also jointly commissioned by BBC Northern Ireland and Radio Telefis Eireann, and sponsored by the Irish Film Board. It was clearly centred on a historical event, the famine of 1848. The way in which it was filmed flirted with the danger of:

> turning abject poverty itself, by handling it in a modish technically perfect way into an object of enjoyment (Benjamin 1973 pp94-5).

as it also emphasized the beauty of landscape, whilst cottage interiors sometimes took on the mellow tones of an old master. The depiction of women fell into the Madonna/Magdalen polarisation, whilst fathers were also shown as emasculated by events both within and beyond their control. However, the occasional deconstructive effects such as voiceovers whilst letters were read, and the use of different brothers to epitomise different discourses such as Ribbon Man, Priest or family man, suggested in a somewhat Brechtian fashion, the slipperiness of history. The presence of the suffering, and indeed starving, body as a bearer of the inscription of colonial might was also to some extent used, so that despite a certain reliance on melodrama character functions and events, the series moved beyond formulaic realism. However, the relatively sympathetic representation of the English Agent, played by Michael Kitchen, muted the critical effect, particularly for British viewers.

Produced by BBC Northern Ireland, single authored plays *The Precious Blood* by Graham Reid and *Love Lies Bleeding* by Ronan Bennet engage in a direct way with the political dimension of the Troubles in a Northern Ireland setting, though largely through the perspective of personal romantic and family relationships caught up within these events, foregrounding the implication of different religious and class contexts within a setting of urban decay and disadvantage. In the former, shown on BBC2 *Screen Two* (June 1996), the plot centres on a cross-community love affair. In the latter, shown on BBC2's themed *Screenplay* season (22 September 1992), the use of flashback as the imprisoned Republican protagonist recalls his girl, now dead, she is typically remembered within a rural setting – a sign both of Irish identity but also of an Eden now out of reach. A recent BBC2 drama, *A Rap at the Door* by Pearse Elliot (07 March 1999), based upon the true story of a Northern Ireland woman who was abducted and never seen again, took the form of three monologues to camera by her three children, Dermot, Cathal and Tierna. Their different perspectives produced an interrogative effect through fragmentation and oddly angled shots, thus hinting at a more fragmented and potentially post-modern approach to the slipperiness of both language and history. The works of Roddy Doyle, including *The Family, The Snapper* and the full-length cinema film *The Commitments* have also all been shown on British TV. Significantly, the urban setting of the former in depressed areas of Dublin, and their gritty yet humourous focus on social problems was not initially well received by all viewers in Ireland. According to Gray and Ryan (1996 p187), there were those who 'refuse to believe the conflict and terror portrayed in *The Family* exists in Ireland'.

Whereas most of the general examples above work broadly within realism, where the relationship with history and politics is closest this approach verges on critical realism. Where the influence of melodrama is strongest, realism is less critical, moving towards formulaic realism. Those pieces which are most critical in their relationship to history and politics are those which are partly driven by humour, and in this closest to the Irish theatre tradition where the comic and the tragic are closely interwoven even today. A realist form combined with a predominantly soap/comedy as in *Ballykissangel* may perhaps inevitably be conservative rather than subversive. It offers a surface reproduction of a generic model rather than a referent, and thus like other 'Simulacra prevail(s) over history' (Baudrillard reprint 1995). Although it has been suggested by Lovell (1982) that sitcoms are least subversive where the reference to social reality is greater, she also comments that disruption may 'degenerate into tiresome and predictable frolics'. Thus there may be potential limitations to the potential radical effect of the absurdist excess of *Father Ted* especially as ironic readings cannot be guaranteed. Nelson (1997) underlines Caughie's point that television may open:

> identity to diversity, and escapes the notion of cultural identity as a fixed volume .... But ...
> it does not do it in that utopia of guaranteed resistance which assumes the progressiveness
> of naturally oppositional readers who will get it in the end (1995, p55).

The most significant differences between *Ballykissangel* and *Father Ted* can be encapsulated in the fact that the latter could not possibly have continued after the death of Dermot Morgan. On the other hand, *Ballykissangel* has survived the loss of the two original main characters, Father Clifford and Assumpta Fitzgerald, the former to another parish, the latter to death, at the end of the third series, when the actors requested write-out. As Rob Brown indicated in an article, the BBC had anticipated this and gradually increased the ensemble qualities of the series, by running some episodes without the stars – as has been the practice with a similar situation in other location centred, bland TV texts such as *Heartbeat*. In the latest 1999 series of *Ballykissangel*, other major characters such as Ambrose have been written out and replaced by fresh characters, further confirming Brown's point that:

> Viewers like the place and the scenery almost as much as the characters. The showsinvari-
> ably have a central character with a job ... the situations come out of the job and can be
> continued... (*Independent* 23 February 1998).

For British viewers then Ballykissangel/Avoca remains a kind of Tir an Og – a land of heart's desire where difficulties are evaded and charming eccentrics live. Despite the job-centred element of *Father Ted* the specific style depended on the protagonist. Further, the latter programme's form is more complex, being both archaic yet avantgarde, involving post-colonial and some post-modern characteristics. The subversive nature of the dionetic laughter prompted by the grotesque and carnivalesque qualities is anti-establishment in a variety of ways which can be related to the post-colonial condition, yet whilst rooted in Irish culture the comedy celebrates post-modern intertextuality and TV technology. However, both texts run the danger of perpetuating for the British audience, condescending stereotypes of Irish otherness, because to different degrees 'History is suspended in a commodified sense of place' (O'Toole 1995 p40). Reception in Britain is also further complicated, for the Irish diasporic population may originate in the North or the South, and thus read differently.

Currently Irish theatre is experiencing a veritable Renaissance-style flowering, with dramas which explore a range of themes across genres and performance styles. It would be especially appropriate since, according to Fintan O'Toole, Ireland, like the Irish, is both everywhere and nowhere, a cultural hybrid. He claims that:

> What contemporary Irish culture is doing in all this is demolishing the colonial opposition of Self and Other and re-inventing the ideal of the Self as Other (1995 p69).

The kind of cross-border and international production collaborations and transmission possibilities indicated above suggest that TV dramatists should seize the present opportunity to create new fluid forms for a new fluid identity. Kiberd has suggested that the British invented Ireland, but perhaps now is the moment for Irish dramatists to create for TV shown in Britain and Ireland, representations that are neither ideal nor grotesque, claiming what Homi Bhabha has called the 'Third Space':

> which constitutes the discursive conditions of enunciation that ensure that the meanings and symbols of culture have no primordial unity or fixity; that even the same signs can be appropriated, translated, rehistoricized and read anew' (1994 p37).

## Notes

1. During the ongoing Peace Process negotiations between the Republicans and Loyalists in Northern Ireland, in which Mo Mowlem for the British Government, and Bertie Ahern the Taoiseach na hEireann have played leading roles with the indigenous politicians, the Good Friday Agreement signed in 1998 has been a significant factor in the reduction of inter-community tensions.

2. See Mary J.Hickman *Religion, Class & Identity: The State, The Catholic Church and the Education of the Irish in Britain* Ashgate 1995 (Paperback 1997).

3. 'We will show that Ireland is not the home of buffoonery and easy sentiment, as it has been represented, but the home of ancient idealism.' Extract from the National Theatre Manifesto of 1897, later to be based at the Abbey Theatre, Dublin, quoted on pp8-9 of Lady Gregory's *Our Irish Theatre*.

4. See Julia Kristeva *Powers of Horror: an Essay in Abjection* Columbia 1982.

5. De Valera's Constitution of 1937 valorised the family unit, and in Article 41 stated, 'woman by her life within the home gives the State a support without which the common good cannot be achieved.' This was challenged by feminists, see Kiberd 1995, p405.

6. See W.B. Yeats play *Cathleen Ni Houlihan*, 1902

## Bibliography

Baudrillard, Jean *Symbolic Exchange and Death* London, Sage, 1993.

Belsey, Catherine *Critical Practice* Methuen New Accents, 1980.

Benjamin, Walter *Understanding Brecht* London, New Left Books, 1973.

Bhabha, Homi *The Location of Culture* London, USA & Canada, 1994 (reprint 1997).

Bowes, Mick 'Only When I Laugh' in A. Goodwin & G.Whannel, *Understanding Television* London & New York, Routledge 1990.

Eagleton, Terry *Heathcliff and the Great Hunger* London, New York, Verso, 1997.

Fiske, John *Television Cultures* London & New York, Routledge, 1987.

Foster, Roy *From Paddy to Mr.Punch* London, Penguin, 1995.

Foucault, Michel 'Of Other Spaces' in *Diacritics* Volume 16, Part 1, Spring 1986.

Gilbert, Helen & Tompkins, Joanne *Post-colonial Drama: Theory Practice, Politics* London & New York, Routledge 1996.

Gledhill, Christine *Home is Where the Heart is: Melodrama and the Woman's Film* London, British Film Institute, 1987.

Gray, Breda & Ryan, Louise 'Gendered Constructions of Irishness' in (ed) S. Stern-Gillet, T Slawek et al *Culture and Identity* Katowice, Poland, University of Silesia, 1996.

Kenneally, Michael (ed) *Irish Literature and Culture* England, Colin Smythe, 1992.

Kiberd, Declan *Inventing Ireland* London, Cape, 1995.

Mercier, Vivian *The Irish Comic Tradition* Oxford University Press, 1962, reprint, London, Souvenir Press, 1991.

Nelson, Robin *TV Drama in Transition: Forms, Values and Cultural Change* London, MacMillan 1997.

Morris, Pam (ed) *The Bakhtin Reader* Britain, Arnold, 1994.

O'Toole, Fintan *Black Hole, Green Card: the Disappearance of Ireland,* Dublin, New Island Books, 1994

Rockett, Kevin, et al *Cinema and Ireland* Croom Helm 1987, London, Routledge, 1998.

Wayne, Mike (ed) *Dissident Voices: the Politics of Television and Cultural Change*, Virginia, Pluto Press, 1998.

# Diagnosing the Alien: *Producing* Identities, American 'Quality' Drama and British Television Culture in the 1990s

Janet McCabe

## Introduction

This chapter aims to examine how American quality dramas are constructed, discursively and institutionally, to produce meaningful identities for and on British television in the late-1990s. Adopting a Foucauldian approach that defines the field of knowledge as inseparable from regimes of power and historically contingent, I will challenge the orthodox paradigm about American programmes by demonstrating how such 'knowledge' is given shape by imagining certain audience identities; and how broadcasting networks function in this process to manage that 'knowledge'. What will be suggested is that meaningful cultural identities for US syndicated dramas, in this case *ER* and *The X-Files*, are produced and anxieties regulated by institutional discourses, or what shall be termed as the 'institutional text,' and made visible by knowledge about an imagined rather than actual viewer. Anxieties are less about fears of being swamped by American imperialism than about how nascent British identities, both culturally and institutionally, are being discursively negotiated.

## British television, 'American' identities

Gone are the days when American acquisitions were considered cheap alternatives to domestically commissioned material on British television. Only recently was Channel 4 involved in a vitriolic price war with Channel 5 over British syndication rights for US hit series such as *ER* and *Friends*. Resolution came when Channel 4 allegedly spent £60 million ($100 million), paid over four years, to the programmes' distributor Warner Bros. to clinch the deal (Smith, 1996, p.20). If British commercial television executives are prepared to pay these exorbitant prices for American programmes, then it follows that they must believe such investments make good economic sense. The logic is that high ratings justify inflated costs which, in turn, are recouped through lucrative advertising revenue. International economic deals have national implications; the resulting situation sparked off further political debate about Channel 4's working practices, remit and plans for privatisation, and the amounts television companies are prepared to pay for American rather than British material. Behind these political machinations lay concerns about national sovereignty and cultural responsibility, and begs the question as to how are we to make sense of American programmes within British television schedules at the end of the 1990s. In short, how is it possible to decipher a television text produced in America as a cultural event experienced in Britain?

American fictional dramas are nothing new to British television schedules. Orthodox critical paradigms view the flow of American television exports into foreign markets as a 'one-way street' (Nordenstreng & Varis, 1974). Informed by neo-Marxist understanding of capitalist culture, these theoretical discourses are replete with accusations about the rampant homogenising forces of the American market, imperialistic forces that are said to dominate world telecommunication systems, threaten national broadcasting sovereignty and render indigenous cultural identities subordinate (Schiller, 1969 & 1991; Silj, 1988, pp.22-58). Statistical evidence about American imports on British television would seem to support such a perilous assessment, as American-made dramas steadily increase their market share in Britain: 30 per cent of dramatic fiction screened on British television originated from the US in 1996, rising to 36 per cent in 1997 and 44 per cent in March 1998 (*Eurofiction*, 1998). Moral panic at these figures has become a site for intense scrutiny, as government-funded organisations gather knowledge about international patterns of communication flow (for example, the Broadcasting Research Unit), trade restrictions legislate against foreign television imports (such as the EU's refusal to allow the telecommunications market to be opened up to free trade under GATT), broadcasting networks institute new delivery systems (communication satellites, cable networks and digital technologies) as well as deregulation, and the new orthodoxy teaches us to resist such totalising theories as that of cultural imperialism.

This chapter aims to challenge the master paradigm still further through an investigation of the discursive processes and institutional discourses that rationalise American 'quality' dramas on British television. Adopting a Foucauldian approach that defines the field of knowledge as inseparable from regimes of power that are historically contingent, this essay will explore the following: how the field of 'knowledge' about American 'quality' drama is constituted; how such 'knowledge' is given shape by imagining certain audience identities; and how broadcasting networks function in this process to manage that 'knowledge.' It is my intention to suggest that meaningful cultural identities for these imported dramas are produced and anxieties regulated by institutional discourses, or what I shall call the 'institutional text', and made visible by knowledge about an imagined rather than an actual viewer. Moral panics about swamping are less about fears of American imperialism than about how nascent British identities, both culturally and institutionally, are being discursively negotiated. I wish to start by making three observations to support my argument.

First, British cultural studies has contributed immeasurably to our understanding of the dynamic social processes involved in the generation and circulation of knowledge within late-capitalist societies, and may help us understand how meaning is produced in dramas that have been removed from their original viewing context and repositioned into a radically different one. The television text is a discursive inter-textual field of contested meanings produced by a capitalist institution, a product of dominant ideology, which, in turn, perpetuates that ideology (summaries, see Fiske, 1987 & 1994). Such analysis underpins current theories on how textual meanings are activated by the television viewer. Central to this work is Stuart Hall's 'preferred reading' theory (Hall, 1980).

Identifying the television text as polysemic, open to several possible readings, he argues that such a discourse is a site of ongoing contestation over meanings, a conflict involving how meanings are encoded by producers, how dominant ideology structures 'preferred' meanings in the text, and how audiences decode. Audience decipherment, dependant on class and social formation, is determined either by aligning with, negotiating or even opposing the 'preferred' meanings at the moment of reception (i.e. American production, British text). What emerges from this 'preferred reading' theory is not so much that different viewers negotiate the same text in different ways but that such knowledge emerges as part of a densely constituted field about modern subjecthood during the 1980s. Such a theoretical conception of an empowered television viewer coincided with the nascent categorisation of modern Britons as meritocratic citizens within specific social practices, both discursive and institutional, and tied to a particular historical moment which, in turn, demanded new cultural institutions to produce representation.

Second, The Peacock Report, calling for changes in broadcasting policy, advocated consumer sovereignty, a consumer able to decide for themselves (*Peacock Report*, 1986). Identifying this 'robust' consumer served political interests, it validated government legislation on deregulation (1988 *White Paper on Broadcasting Policy*; 1990 *Broadcasting Act*), cleared the ground for financial and managerial restructuring at the BBC, justified a competitive free-market and audience targeting, and made possible market-driven communication satellite and cable network services, increasing the number of channels available and creating numerous specialised ones aimed at different audiences. Given a Foucauldian understanding of discourse there is no reason other than to suggest that this 'consumer' is a discursive formation. A specific socio-historic individual that is interpreted not only as a construct, but as an effect of social practices through which its very identity and desires are formed and constituted; put another way, the 'robust' consumer is a discursive and institutional category. Each television viewer is constituted as subject in, and subject to, what I call the 'institutional text'. The 'institutional text,' as a function of an individual broadcasting network, produces meaningful identities (in this case, identities from American-imported dramas) by mobilising and managing specific knowledge about its imagined viewer. Such knowledge is held in place by a densely constituted set of values and assumptions which underpins network identity, and the means by which stations justify and regulate themselves, be it through legislative measures, broadcasting remits, scheduling policy or ancillary texts such as printed media or Internet sites. Out of this knowledge emerges not only a viewer shaped to the institutional requirements of the network, but one that in a sense makes cultural identities visible.

Third, anxieties about cultural imperialism make television liable to intervention; regulatory regimes protect the institutional text which, in turn, embodies those discourses. Further investigation into British broadcasting practice, for example, challenges the notion of exactly how far American programming actually dominates British television. Seldom have American dramas topped the British ratings, despite alarmist speculation to the contrary. British television's retention of regulatory regimes, in spite of deregulation and changes to public service broadcasting (*Home Office*, 1987; 1988), obliges us to recognise broadcasting's own capacity to sustain its own indigenous market. Regulatory intervention is enforced through a scheduling policy that reserves prime-time slots for mostly domestically commissioned material, for example, while 39 per cent of television dramas shown on British terrestrial channels in October 1998 originated from America, with 45 per cent from the UK, only 11 per cent were scheduled in prime-time slots as opposed to 89 per cent devoted to British-made dramas; again in March of that year, even though 44 per cent of dramatic fiction was acquired from America, compared with 40 per cent made in Britain, the prime-time statistics remained the same as above (*Eurofiction*, 1998). No more than 15% of overseas imports can be shown at any one time on terrestrial networks which adds further credence to the argument that the possibilities for seeing and identifying with American representations are diligently regulated. Even though two million additional viewers tuned in when the BBC rescheduled *X-Files* from BBC2 to BBC1, access to American drama on terrestrial television is controlled and rationed by regulatory practices, practices that are enforced by regulatory agencies such as government sanctions, the Independent Television Commission (ITC) and censoring bodies (for example the Broadcasting Standards Council). Broadcasting policies, policed by government legislation and control over funding, illustrate how institutional regimes protect British television's own communication sovereignty and notions of national identity; and how such procedures regulate *representations*, be they home-produced or transnational, on UK television to create an 'imagined community' (Anderson, 1983).

I should like to focus on two of the most interesting American 'quality' dramas to emerge in recent years. *ER*, NBC's top-rating medical drama series acquired by Channel 4 in 1994 and *The X-Files*, FOX's quality drama-turned-cult hit which first aired on BBC2 in the same year before being transferred midway through season two to BBC1. American 'quality' drama emerged, made possible by legislative and institutional changes, on US commercial prime-time television in the early 1980s, and has now become a small yet viable part of the network schedule. Difficult to define, these one-hour dramas constituted a new brand of sophisticated programming aimed directly at a young upwardly-mobile audience (Thompson, 1996). Crucial as these quality dramas may be to the fate of television culture at the end of 1990s and beyond, this is secondary to the arguments I intend to detail here. My contention is that meaningful cultural identities for these dramas are constituted and organised by an institutional text, produced by a particular British broadcasting network, and one which operates through a dense field of text and techniques. Such discursive processes are dependent above all on an imagined viewer-subject to produce knowledge which, in turn, sustains corporate hegemony. In the case of Channel 4, the *ER* viewer is subject to, and subject in, an institutional text that gives shape to the network's hegemonic cultural production. By contrast, *The X-Files'* viewer produced by the BBC institutional text makes visible a new type of television viewer, one that points to a crisis in public service broadcasting, and one that reveals the difficult choices facing the Corporation in its medium-term future. If new cultural identities emerge, then it does not occur independently of the remaking of the viewer-subject, for it is through this imagined viewer that identities are made visible.

## Our medicinal friends: Channel 4 and *ER*

Reliance on American syndicated material has helped Channel 4 build its corporate identity. *ER*, America's top-rating 'quality' drama series, set in a fictional Chicago-based hospital, with an average weekly audience of 33 million in the USA, is one such purchase. In contrast to the mainstream television exposure NBC's prime-time drama receives in America, *ER* is televised on the small independent terrestrial network, attracting a respectable four million viewers for its British syndication channel. Acquisition of such dramas from the US, programmes defined as literary-based, generically groundbreaking, tending towards liberal humanism, and made by those with a 'quality' reputation (Thompson, 1996, pp.13-16) for 'yuppie, TV-literate baby boomers' (Feuer, 1995, p.8), sits well with Channel 4's remit that encourages diversity and experimentation, and a market position that allows it to be innovative. Far from merely copying lessons learnt from the American networks, Channel 4 has bought in these dramas to serve the interests of its own network identity. Channel 4's remit, a mandate which professes a commitment to 'fresh and invigorating television that challenges the norm' (*Channel 4*, 1998, p.1), and aspires to be the brand choice for 'modern culture in Britain' (ibid.), generates a new set of cultural meanings for *ER*. Through re-contextualising this popular mainstream American drama, Channel 4 has produced an institutional text which defines 'quality' in terms of distinctiveness, intelligence and innovation precisely by mobilising knowledge about its imagined viewers, viewers that are profiled for being 'independently-minded' (Willis, 1996, p.22), and ones that justify policy decisions and guide programme acquisition; as John Willis, Director of Programmes for Channel 4, recently said: 'We have a responsibility to our viewers to get good acquisitions' (Smith, 1996, p.20).

New cultural identities, emerging in the 1980s, revealed political representations that the party system had not yet evolved to reflect. How gender and ethnicity were proportional to traditional class hierarchies that were in the process of breaking-up and how competing histories made visible by the traditional 'Other' (women, the working-class, and regional and ethnic groups) had reinterpreted and re-conceptualised the model of what the British nation was. Existing cultural and political institutions were proving incapable of producing adequate representation when Channel 4 began transmission in 1982, as a product of governmental legislation that required it to appeal to those not generally catered for by ITV. Turning now to Channel 4's syndicated US drama *ER*, televised over a decade later, I will examine how the institutional text generates meaning for this show by mobilising 'knowledge' about an imagined audience, one that is both an historical subject and the site of certain institutional demands and procedures of subjectification. The institutional text, on an extra-diegetic level, asserts a particular aspirational lifestyle group to give meaning to the drama's dense visual and narrative style, and on a diegetic level, and at the same time, constructs a Utopian image of nationhood. In short, the institutional text, through asserting knowledge about its imagined viewers, negotiates nascent identities in 1990s Britain while shaping representation to the requirements of its own institutional identity.

*ER* was initially publicised in Britain as bringing Hollywood talent to the small screen. Much was made of *ER*'s pedigree credentials in the shape of Michael 'man of the moment' (Ogle, 1995, p.17) Crichton and Steven Spielberg before the drama first aired on Wednesday nights at 10pm. Almost all the quality newspapers previewed the drama by recounting Crichton's ten-year quest to have his script, based on experiences as a Harvard-educated medical intern at Massachusetts General Hospital, made into a film, and the participation of Hollywood's most commercially successful director, Steven Spielberg, who had just picked up several Oscars for *Schindler's List*, and whose production company Amblin made *ER*. Authoring the drama gave meaning to the multi-layered narrative and fast visual pace: 'All Crichton's thrillers from *Jurrasic Park* to *Disclosure*, have been about people and workplaces under pressure: he is a master of information overload, who can make you believe in crises' (Purves, 1995, p.5); and of the Spielberg connection: 'the piece has a claustrophobic intensity which befits its status as a Steven Spielberg production' (Barnard, 1995, p.47). A glance at Channel 4's press releases reveal the source for such a campaign, for it built a marketing strategy around past Crichton/Spielberg associations and on their ability to elevate a national television drama to the level of international blockbusting cinematic entertainment. Prepublicity branding only makes sense by rendering visible and defining a new kind of viewer-consumer, one that is, on the one hand, able to recognise marks of authorship akin to an independent art-house spectator and, on the other, demands a unique televisual experience; that is, exclusive designer-label television for *cineastes*.

Channel 4's institutional text continues to understand *ER*'s big budgets, high production values and 'cinematic' style: the scripting of plot-dense narratives by a top notch writing team, shooting on film before transferring to videotape, non-traditional lighting systems (overhead fluorescent bulbs), camera style (the state-of-the-art Steadicam) and fast editing (700-800 cuts as opposed to the standard 300-400 edits for a one-hour drama) (Oppenheimer, 1995) by asserting the *cineaste* viewer interested in quality and designer labels. Such a viewer is spoken about as being able to keep pace with the narrative verisimilitude and exhaustive sensory experience: 'the viewer's attention is not allowed to flag for a moment' (Massingbred, 1995, p.29). This discourse elevates *ER* to the status of 'art film' by evoking a viewer fitted to the task of spectacular consumption and the demands of home cinema entertainment which, in turn, justifies Channel 4's substantial investment in the drama, and assures the station's cultural kudos as an innovative network: 'That means a channel willing to try the new in both form and content' (Willis, 1996, p.22).

Two episodes stand out as Channel 4, with its reputation for sponsoring innovative independent films, reclaimed *ER* as its own through activating new associations. Quentin Tarantino's directorial effort in season one, and Ewan McGregor's guest appearance in the season three drugstore armed-robbery episode. Both episodes were uniquely positioned as small experimental 'movies'. Tarantino's cult reputation as an independent film *auteur* meant that this episode was promoted as 'must see TV': 'Tarantino's direction ... took the programme into a different league. It was the best instalment of *ER* I have ever seen' (Pile, 1995, p.12). Knowledge, such as the scuffle between two gang girls, one holding a severed ear, or Dr Peter Benton's (played by Eriq la Salle) graphic handling of the bone saw, is unlocked by those who could understand the episode's *auteur* underpinning: 'Last night, the series must have doubled its rating with all the movie nerds who were watching just to check Tarantino's *auteur* touch in the surgery scenes' (Romney, 1995, p.11). Ewan McGregor's guest appearance inspired similar copy: 'Cinema's coolest junkie visits the small screen's hippest hospital' read *The Observers'* headline review. McGregor, critically-acclaimed star of contemporary British cult films such as *Shallow Grave* and *Trainspotting*, is strongly identified with Channel 4's *FilmFour* trendy lifestyle-choice corporate image. Both episodes, plucked from their original narrative contexts, were repeated together early in 1999, reasserting the distinctiveness of these tele-cine experiences and bolstering Channel 4's reputation as an investor of innovative, independent film production. That both celebrities were recognised as being 'big' *ER* fans, and that their star images were promoted as lending originality to the *ER* text leads me to conclude that the insitutional text produces an identity for *ER* as innovative, generically ground-breaking and stylised precisely by bringing such high-profile admirers into its own discourse. Aligning viewers with other *ER* fans like McGregor and Tarantino is less about the series' popularity than a 'concept' of *ER,* and a viewer identity defined by lifestyle choice, attitude and fantasies of self.

Further evidence to support this reading is provided by 'knowledge' about the typical Channel 4 viewer:

> ... young or old, rich or poor, Channel 4 viewers are independent-minded. They expect curiosity, challenge, controversy. They will watch Italian soccer and Cezanne, current affairs and Chris Evans. Above all, they don't want easy answers but to make sense of the world themselves (John Willis quoted in Smith, 1996, p.22).

The statement proffers a model of the average Channel 4 viewer as an autonomous individual able to make choices for themselves, and one that plunders different cultural identities and lifestyle alternatives to suit their own unique sense of self. What is interesting about this conception of the Channel 4 viewer that I am about to analyse in relation to *ER's* narrative is that audiences are invited to construct their own identity from a range of multicultural, national and transnational identities positioned within the television 'flow'. Dispossession from traditional familial and cultural roots means that Channel 4's 'imagined community' is one in which aspirational social identities have become a matter of pluralist personal creeds, self-exploration, lifestyle choices and new cultural needs.

American dramas instituted as British texts offer alternative representation for nascent British identities. Channel 4's institutional text positions *ER* as offering an image of the Utopian hetereogenic community within its narrative diegesis, one that sets out to forge working unity out of widely differing immigrant, gender and cultural experiences away from a recognisable social world. Despite personal tensions, the medical team functions cohesively under pressure because, rather than in spite, difference. Even the waiting room is composed of the right 'ethnic' mix (Cassidy, 1997, p.7) for an urban trauma centre. At a time when the notion of nationhood is the subject of furious public debate, the institutional text produces a different kind of nation, one that retains particular ideological aspects of nationhood (social responsibility, professional ethics, and a meritocratic social order) while losing the more oppressive aspects of Englishness based on social hierarchies and racial and gender prejudice.

> The Americans offer brash newness ... they have speed, wit in clipped dialogue, accents which to English ears carry few of the graduated social distinctions that lumber popular drama over here (Gritten, 1995).

The institutional text thus reproduces a Utopian variation of the nation, one that is more palatable to Channel 4's network identity, and one that is made sense of by evoking an aspirational social group.

Behind these identities, where individuals feel they can make choices about their national identity, lie anxieties about nationhood. Questions about national identities are once again plaguing political discourse, constitutional reform and the formation of new political institutions. As devolution changes the political landscape and a decision on whether or not Britain should embrace European integration draws ever closer, the notion of what it means to be British, and perhaps more importantly English, is being redefined. Concerns about alleged prejudice against Englishness from conservatives are equally matched by anxieties from liberals about how it is tainted by racism and class distinctions. Channel 4 has instituted resistance as part of its network identity (Channel 4's alternative 1997 Christmas message from the parents of murdered black teenager, Stephen Lawrence provides evidence of such dissent). It produces troubled national identities, identities that pose difficult questions about sovereignty, offering few easy answers, but which nonetheless relish in the confusion and potential for change. American identities, then, are positioned by the institutional text to negotiate nascent identities as well as to regulate that process through displacement onto the cultural 'Other'.

American hegemony that seeks to forge national consensus from different social groups is repositioned to offer dissent identities in Britain, a repositioning which is made visible by an imagined community of viewers seeking alternative representation. I wish to contend that Channel 4's institutional text, one that embraces multicultural diversity, gender equality and social aspiration, buys in material to strengthen its discourse. On one level, the high-energy visual pace, mixing visceral cinematic spectacle with the sensationalism of 'real-life' television and news coverage, underlines Channel 4's unique commitment to innovative independent film-making and lifestyle consumer choice; on another level, the diegetic narrative culturally constructs nascent identities. Meanings come into being by mobilising knowledge about aspirational social groups, groups that not only make sense of the narrative trajectories but are described as a precondition for network identity.

## Alien encounters: BBC and *The X-Files*

Unlike Channel 4's handling of these imported series, American 'quality' dramas occupy a more uncomfortable position in the BBC's schedules, seeming more out of keeping with the Corporation's hegemonic cultural production, and often confined to the art-house remit of BBC2. *The X-Files*, FOX's cult-turned-international phenomenon about two FBI agents investigating unsolved cases involving paranormal occurrences and government conspiracies, is such an example. Far from denigrating *The X-Files* to the margins for minority audiences, its acquisition reveals much about the BBC's corporate identity at this time. While the modern public arena is negotiated, both diegetically and extra-diegetically, in terms of social dislocation and identity confusion linked to state institutions and technology, such an experience speaks of yet another public crisis, namely the assault on public service broadcasting and the BBC remit. It is an agenda that contends with questions of how best to serve public interests, a struggle about controlling agency, conflicting interests and ideas, between paternal guardianship and public choice, 'internal diversity' (mixed programming) and 'external diversity' (network branding) (Collins, 1992, pp.94-98), and the problem of how best to absorb social heterogeneity and reduce ethnic difference into a dominant idea of the 'public'. It is a debate that remains unresolved, and one that finds the viewer at its centre.

Knowledge about the imagined BBC viewer is implicit in the 1986 *Peacock Report*. Embedded in the Committee's recommendations about how the new BBC remit should balance commercial and public service enterprise is a particular type of viewer:

> Our own conclusion is that British broadcasting should move towards a sophisticated system based on *consumer sovereignty*. That is a system which recognises that viewers and listeners are the best ultimate judges of their own interests, which they can best satisfy if they have the option of purchasing the broadcasting services they require from as many alternative sources of supply as possible. There will always be a need to supplement the direct consumer market by public finance for programmes of a public service kind ... supported by people in their capacity as *citizens* and *voters* but unlikely to be commercially self-supporting in the view of broadcasting entrepreneurs (Peacock, 1986, paras. 592, 593.) [My emphasis].

This viewer, able to make informed choices for themselves, both as a consumer and citizen, is a far cry from the 'vulnerable' BBC viewer identified by Pilkington in 1968. Pilkington defined this viewer as an uncritical consumer in need of platonic protection, and one which justified the Reithian model of public service broadcasting (Pilkington, 1968). Couched within Pilkington's rhetoric lay class ridden assumptions about nationhood, a nation divided between those in need of guidance (such as the working-class mass) and those able to make the appropriate programming decisions on their behalf (the educated middle-class elite). Thirty years on, the case for preserving such a paternalistic model is less than clear. Difficult choices facing the Corporation are inseparable from the confusion over what exactly 'public' means within modern Britain (especially since social and ethnic diversity resists easy assimilation into a dominant notion of the 'public'), how such debates have been further re-configured by new technologies and competing channels such as Channel 4, and how the imagined BBC viewer has been transformed in the process. Looking now at the BBC's transmission of *The X-Files*, what I hope to achieve in this last section is to discuss how Pilkington's model of the vulnerable viewer has collapsed, replaced by a completely different idea of what the television viewer is, as described by Peacock. I wish to contend that the institutional text negotiates issues of the 'public' in *The X-Files* by mobilising knowledge about a new type of modern viewer. On one level, this viewer gives meaning to how the popular text is consumed and its multivalent narrative understood. On another level, and at the same time, such knowledge allows for a questioning about the very notion of the public/private in terms of identity, spaces and the state. Asserting the imagined viewer, and I will go on to describe that particular type of viewer next, the BBC institutional text negotiates anxieties about modern public identities, both discursive and institutional, while asserting that crisis in relation to its own corporate future and public service broadcasting remit.

Unlike Channel 4's prepublicity campaign for *ER*, *The X-Files* received little attention when it first aired after the watershed on the Corporation's minority channel BBC2. It soon became a hit, attracting audiences in excess of 5.5 million at the beginning of January 1995: '*The X-Files* was allowed to find its cult audience on BBC2 with little attendant hullabaloo' (Anthony, 1996, p.18). Even though it became BBC2's top-rated series, a scheduled episode was dropped in favour of a one-off opera event, a programme which didn't even figure in the top-30 most watched shows for that week. Moving to BBC1 midway through season two meant an additional two million viewers, pushing audience figures up to nearly ten million per episode. Despite this, the series is prone to cancellation, rescheduling and a variable time slot. Such scheduling is justified by mobilising knowledge about a fanatical fan base, eager viewers who scan the schedules, and for whom the series is an unmissable event. Widespread media preoccupation with audience loyalty offers us an ever more precise description of a new kind of devoted-viewer, a consumer of the televisual and other related texts (i.e. videos, books and related official merchandise franchised by FOX), a narrative agent who has found a new electronic public space to dissect the storylines, and an avid follower who justify scheduling policy. Far from suggesting that this type of viewer is the only viewer, I contend that knowledge about this devoted viewer, and one that enjoys a high media profile, has been mobilised to make sense of the 'public', both narratively and institutionally. In short, it shows how the idea of this 'imagined community' is being shaped to the requirements of new technologies, television programming and within contentious debates about the 'public,' both discursively and institutionally.

If *The X-Files* narratives speak of governments controlling human destiny (viruses, genetic engineering) and identities (alien DNA), the devoted-viewer is institutionalised as having gained access to that information through repeated television watching, purchasing of ancillary texts and/or being part of a computer-mediated communications network with other hard-core fans (the two official Web-sites boasts around one million hits per month). Devotees or 'X-philes' as they are more commonly known use cyberspace to exchange information, ponder over minutiae and dissect the multi-layered narrative. Accumulating information in this way affords individual users the opportunity to offer opinion and even subvert narrative agency. Narratively, Fox Mulder (David Duchovny) and Dana Scully (Gillian Anderson) are continually seeking explanations related to personal family histories or government cover-ups (the black oil). Scully's rationalism, shaken by her abduction, terminal cancer and genetically-hybrid child Emily, and Mulder's quest to find answers to his sister's abduction, possibly by aliens, has rendered his parentage questionable (the revelation that Cancer Man, played by

William B. Davis, might be the real father of Samantha, his sister, and of himself, complicates this storyline still further) both reveal that narrative agency should be treated with deep suspicion or, as Mulder's log-on reads, 'TRUSTNO1'. Narrative agency is given an unprecedented mobility and exchangeability as a consequence. More importantly though, it is the status given to the denizens of cyberspace to interpret these narrative strands which allows such an issue in a sense to become visible. The devoted viewer, as a consumer of multiple texts and television citizen whose voice is heard and opinions sought, defines ever more precisely the viewers as a community of individuals bonded by exclusive knowledge and designer-label products. What occurs is a new configuration of the public being about shared ownership and investment, both financial and personal, a concept that avoids defining them in terms of national, social or ethnic groupings: 'It is one of those rare series that we have all discovered for ourselves, and because of that we cherish it all the more' (Berkmann, 1995, p.55).

Fostering knowledge about the virtual community and ownership is dependant on a particular re-conceptualisation of the 'public' space, redefining the relation between the individual and external social world. Diegetically, the public exterior world is a vulnerable space; political surveillance, government conspiracies and unexplained phenomena render the public arena a dark and sinister place, a sense of permanent insecurity, paranoia and fatalism pervades the drama. Above all it indicates the emergence of new models of subjectivity. First of all Mulder and Scully's reliance on portable technology, such as cellular phones and laptop computers, perform an operation of individuation; that is, the technology defines the individual as isolated, autonomous and dispossessed from traditional familial and cultural roots. But it also impels a kind of meditative withdrawal from the public exterior world, in order to allow for experience to be authenticated and legitimised. Technology thus shapes a reclamation of the public arena; and the cyberspace community are integral to that process. While public life is increasingly defined as being about political disenfranchisement and powerlessness (and I am thinking here about electoral apathy and that more British people play the national lottery than voted in the last general election), the devoted viewer, and replicating other forms of confessional television, is positioned as recuperating the public arena through creating their own private meanings, sharing their own experiences and telling their own personal stories. If what it is to be modern is played out within the public arena, both narratively and extra-diegetically, it does not occur separately of the remaking of the viewer-subject, shaped to the requirements of new technologies.

*The X-Files* picked up the People's Award at the 1996 BAFTA ceremony, beating the BBC's high profile period drama *Pride and Prejudice* into second place, and astounding BBC pundits: 'It was a case of urban youth trouncing suburban middle-aged England, of voting via the Internet taking over from the good old stamped envelope' (Brown, 1996, p.15). First of all the award institutes viewers' ownership, despite its outside sponsor. But the ground swell of popularity for the US series uncomfortably demonstrated the conflicting interests between public service and commercial broadcasting. *Pride and Prejudice* was, after all, the jewel in BBC's 1995 scheduling crown; not only this, but the American import, a series that initially made little impact in the States and so was a cheap acquisition for the BBC, had thrashed the Corporation's other lavish costume drama *Martin Chuzzlewit* in the ratings. In so far as Cancer Man and his shadowy cohorts are under threat for standing between us and the truth as Mulder and Scully edge ever closer to unravelling the grand conspiracy, the great and good at the BBC, those who have traditionally stood as influential agents between the masses and 'quality' broadcasting (and 'quality' is still seen as key to the Corporation's brand identity), and their abilities to adequately represent the 'public' and define public service broadcasting appeared to be in question. Couched beneath media speculation about cultural rebellion and moribund representatives lie fears about the selections that the public would make if left to choose for themselves, a dilemma that has been made visible by the devoted viewer, blinded by brand loyalty, eager to participate, who, in turn, sustains that discourse.

As the Corporation in the late 1990s defends itself against charges of 'dumbing down', by claiming that it has a duty to provide programmes that appeal to all, the BBC's institutional text for *The X-Files* provides evidence for how the network is redefining the popular. BBC scheduling institutes *The X-Files* as a weekly uninterrupted 45 minute episodic drama series, a marked contrast to Sky's high-profile presentation of the show, complete with numerous advertisement breaks, programme sponsorship and repeats. Scheduling episodes as unbroken dramatic narratives has meant that the BBC have produced an alternative identity for *The X-Files*: an innovative quality-cum-formulaic dramatic text, defined by high production values (associated with Hollywood, but filmed in Vancouver so not tainted by the mass-produced entertainment label), timely subject matter (millennium agitation and speculative science), narrative structure that shifts between episodic and serial, well-crafted scripts and characterisations. Generating these meanings for the series, the BBC reshapes the popular formulaic text in terms of its quest for a quality tele-literate audience. Niche audiences are nothing new, already associated with 'cult TV'. Much more than a television geek though, the devoted viewer is an active agent in the shaping contemporary television culture. In a sense, the nerd has evolved into an arbiter of quality programming, and thus allows the show to be absorbed into a quality-cum-formulaic programming remit.

Unlike Channel 4 and its aspirational audience, the BBC, on the verge of change, finds a consensual notion of the public more difficult to sustain, partly because British society can no longer be understood in terms of the dominant class paradigm, if at all. Contradiction is built into the BBC institutional text as it integrates the dramas appeal into its own remit. Defining *The X-Files* as a quality-cum-formulaic product, through evoking the devoted viewer, justifies new forms of cult programming that will hook in viewers at a time of increased competition amongst individual networks with a strong brand image (Channel 4), but bolsters the BBC's reputation for investing in original *Zeitgeist* television (the drama's estimated worldwide audience stands at around 100 million (Bows, 1997)).

## Conclusion

If, as John Reith, first Director-General of the BBC, contests, a nation's broadcasting reflects a nation's conscience, then the question that comes to mind is what are the processes that make a national consciences visible and rendered knowable? What I have tried to give a sense of in this chapter is how various institutional texts produce identities by mobilising knowledge about an imagined audience. Any in-depth investigation into either series demands far more attention than I can give it within the word limit. My purpose for gesturing towards these two American 'quality' dramas has been to indicate how these texts have been reconstituted by the institutional text to negotiate how the attributes of modern citizenries in Britain are being shaped and nascent identities constituted. The more the newly empowered television viewer has been described as asserting their own identities, then the more the issue of identity has itself become a locus of knowledge and object of intense investigation. Not least in a climate of global telecommunication expansion where American 'quality' television dramas are increasingly becoming both a subject for regulation and a condition, as in the case of Channel 4 and other cable channels, for corporate survival, how American television dramas are being borrowed to negotiate nascent British identities continues to be an issue worth exploring.

## Acknowledgement
I would like to thank Bruce Carson and K. J. Donnelly for much needed advice in the early stages of this project. I also owe a debt of gratitude to Peter Kramer for his critical insights, and to Mike Allen for all his continued support.

## Bibliography

Allen, R.C. (ed), *Channels of Discourse: Television and Contemporary Criticism*, London, Methuen, 1987.

Allen, R. C. (ed), pp 284-326.

Anderson, B., *Imagined Communities: Reflections on the Origins and Spread of Nationalism*, London, Verso,1983.

*Channel 4: New Developments and Future Ambitions*, London, Channel 4, 1998.

Collins, R., *Television: Policy and Culture*, London, Unwin Hyman Ltd, 1990.

*Eurofiction 1998*, Fondazione Hypercamp Osservatorio sulla Fiction Italiana, 1998 (unpublished).

Fiske, J,. *Television Culture*, London, Routledge, 1994.

Fiske, J., 'British Cultural Studies and Television' in *Channels of Discourse: Television and Contemporary Criticism*.

Feuer, J., *Seeing Through the Eighties: Television and Reaganism*, London, BFI, 1995.

Foucault, M., *Discipline and Punish: The Birth of the Prison*, London, Penguin, 1977.

Home Office, *Broadcasting in the '90s: Competition, Choice and Quality*, Cm 517, London, HMSO, 1988.

Home Office, *Subscription Television*, London, HMSO, 1987.

Jenkins, H., *Textual Poachers: Television Fans and Participatory Culture*, New York, Routledge, 1992.

Lavery, D., Hague., A & Cartwright, M., (eds) *Deny All Knowledge: Reading the X-Files*, London, Faber and Faber, 1996.

Nordenstreng, K., and Varis, T., *Television Traffic: A One-Way Street?* Paris, UNESCO, 1974.

O'Malley,T., *Closedown? The BBC and Government Broadcasting Policy, 1979-92*, London, Pluto, 1994.

Peacock, A., *Report of the Committee on Financing the BBC*, Comnd 9824 , London, HMSO, 1986.

Schiller, H., *Mass Communications and the American Empire*, New York, Augustus M. Kelley, 1969.

Schiller, H., 'Not Yet the Post-Imperialist Era,' *Critical Studies in Mass Communication*, 8, 1991, pp 13-28.

Silj, A., *East of Dallas: The European Challenge to American Television*, London, British Film Institute, 1988.

Silj, A., *The New Television in Europe*, London, John Libbey, 1992.

Sinclair, J., Jacka, E., and Cunningham, S. (eds) *New Patterns in Global Television: Peripheral Vision*, Oxford, Oxford University Press, 1996.

Smith, C., 'Friendly Rivals' *Broadcast*, 20 December 1996, p.20.

Thompson, R.J., *Television's Second Golden Age: From* Hill Street Blues *to* ER, Syracuse University Press, 1996.

Thorpe, V., 'It's all down to the Vision Thing,' *The Observer*, 23 May 1999, pp18-19.

Willis, J., 'The Attitude Channel,' *Broadcast*, 20 December 1996, p 22.

## Newspapers and articles: *ER*

Barnard, P., *The Times*, 2 February 1995, p 47.

Brooks, R., TV Times, *Observer*, 13 April 1997, p 10.

Cassidy, J., and Taylor, D., 'Doctor, doctor, where can I get an aspirin' *The Guardian*, 12 December 1997, pp 6-7.

Clarke, S,. 'A Medical Drama to Quicken the Pulse,' *Daily Telegraph: TV & Radio*, 28 January 1995, p 3.

Gritten, D., 'The Show that set a Nation's Pulse Racing' *Daily Telegraph*, 23 October 1997, p 26.

Gritten, D., 'Why They Do It Better,' *Daily Telegraph*, July 1995.

Joseph, J., *The Times*, 8 November 1997, p 43.

McMahon, B., 'Pure Theatre as *ER* Goes Live,' *Evening Standard*, 18 September 1997, p 15.

Massingbred, H., 'Healthy Diagnosis for Medical Drama,' *Daily Telegraph*, 16 February 1995, p 29.

Odone, C., A Medical Drama in Perfect Health, *Daily Telegraph*, 15 May 1997, p 38.

Ogle, T., 'Forceps Saga' *Time Out*, 1 February 1995, p 17.

Oppenheimer, J., 'Diagnosing *ER*'s Practical Approach,' *American Cinematographer*, October 1995, pp 46-52.

Parker, I., 'What a Bunch of Drips' *The Observer*, 26 October 1997, p 8.

Pile, S., *Daily Telegraph*, 15 July 1995, p 12.

Purves, L., 'Supercharged, Supercool, Superior,' *The Times*, 28 January 1995, p 5.

Sweeting, A., *The Guardian*, 9 February 1995, p 9.

## Newspapers and Articles: *The X-Files*

Anthony, A., *The Observer Review*, 14 January 1996, p 18.

Berkmann, M., 'Excellent Files Bow Out,' *Daily Mail*, 17 March 1995, p 55.

Bows, B., 'UFO-ric Return' *Time Out*, 24 September – 1 October 1997.

Boyd, A., *The Guardian*, 9 February 1996, p 2.

Brown, M., 'Sci-Fi is the People's Choice,' *Daily Telegraph*, 27 April 1996, p 15.

Lyttle, J., 'Do We Need *The X-Files*,' *The Independent*, 6 May 1996, p 8.

McLean, A., 'Media Effects: Marshall McLuhan, Television Culture and *The X-Files*,' *Film Quarterly*, 51:4, Summer 1998, pp 2-9.

Rawsthorn, A., 'Video Sales Zoom as Fans Snub Rental,' *Financial Times*, 13 May 1995.

Rowland, G., 'Waiting for the Third Coming,' *The Observer Review*, 3 March 1996, p 4.

# Index